D.H. Lawrence and Survival

D.H. Lawrence and Survival

Darwinism in the Fiction of the Transitional Period

RONALD GRANOFSKY

McGill-Queen's University Press
Montreal & Kingston · London · Ithaca

ISBN 0-7735-2544-0

Legal deposit second quarter 2003
Bibliothèque nationale du Québec

Printed in Canada on acid-free paper that is 100% ancient forest free (100% post-consumer recycled), processed chlorine free.

This book has been published with the help of a grant from the Humanities and Social Sciences Federation of Canada, using funds provided by the Social Sciences and Humanities Research Council of Canada.

McGill-Queen's University Press acknowledges the support of the Canada Council for the Arts for our publishing program. We also acknowledge the financial support of the Government of Canada through the Book Publishing Industry Development Program (BPIDP) for our publishing activities.

National Library of Canada Cataloguing in Publication

Granofsky, Ronald, 1950–
D.H. Lawrence and survival: Darwinism in the fiction of the transitional period / Ronald Granofsky.
Includes bibliographical references and index.
ISBN 0-7735-2544-0
1. Lawrence, D.H. (David Herbert), 1885–1930 – Knowledge – Evolution (Biology) 2. Survival in literature. 3. Darwin, Charles, 1809–1882 – Influence. 4. Lawrence, D.H. (David Herbert), 1885–1930 – Criticism and interpretation. I. Title.
PR6023.A93Z63 2003 823'.912 C2002-905868-6

Typeset in Palatino 10/12
by Caractéra inc., Quebec City

To Birgitte, Thomas, and Jonas

Contents

Acknowledgments

I owe a debt of thanks to Birgitte Granofsky, with whom I discussed countless aspects of this project as it took shape and whose questions and suggestions about the manuscript were a valuable contribution to what finally emerged. I would also like to pay tribute to the generosity of my colleagues and friends Michael Ross and Joe Adamson, who read this work in manuscript and made many helpful suggestions. Their interest and encouragement during the submission process was heartening. Thanks also to my research assistant, Ken Paradis, and to the readers of the manuscript, one for McGill-Queen's University Press and one for the Humanities and Social Sciences Federation of Canada, who gave generously of their time and criticized constructively. A warm thanks is due as well to Michelle Weinroth for shepherding my manuscript through the intricacies of the Federation's Aid to Scholarly Publications Programme and for keeping me so well informed electronically of the progress and delays. The editorial staff at McGill-Queen's University Press were delightful to work with, in particular the executive director and editor, Philip Cercone, coordinating editor Joan McGilvray, and copy-editor Lesley Barry.

Earlier versions of some of the material in this book appeared in print as the following: "A Second Caveat: D.H. Lawrence's *The Fox*," *English Studies in Canada* 14 (1988): 49–63; "Illness and Wellness in

D.H. Lawrence's *The Ladybird," Orbis Litterarum* 51.2 (1996): 99–117; "Survival of the Fittest in Lawrence's *The Captain's Doll," D.H. Lawrence Review* 27 (1997–98): 27–46; "'Jews of the Wrong Sort': D.H. Lawrence and Race," *Journal of Modern Literature* 23.2 (1999–2000): 209–23; and "Modernism and D.H. Lawrence: Spatial Form and Selfhood in *Aaron's Rod," English Studies in Canada* 26 (2000): 29–51. I would like to thank the editors of these periodicals for permission to use this material.

Abbreviations

The following is a list of the abbreviations used in parenthetical references to Lawrence's works. For full bibliographical details, see Works Cited.

AR	*Aaron's Rod*
CSS	*The Complete Short Stories of D.H. Lawrence,* Volume 1
EmyE	*England, My England and Other Stories*
Fox	*The Fox, The Captain's Doll, The Ladybird*
FU	*Fantasia of the Unconscious and Psychoanalysis and the Unconscious*
L i	*The Letters of D.H. Lawrence: Volume I*
L ii	*The Letters of D.H. Lawrence: Volume II*
L iii	*The Letters of D.H. Lawrence: Volume III*
L iv	*The Letters of D.H. Lawrence: Volume IV*
L vi	*The Letters of D.H. Lawrence: Volume VI*
L vii	*The Letters of D.H. Lawrence: Volume VII*
LG	*The Lost Girl*
MEH	*Movements in European History*
MM	*Mornings in Mexico and Etruscan Places*
MN	*Mr Noon*
P	*Phoenix: The Posthumous Papers of D.H. Lawrence*
PII	*Phoenix II: Uncollected, Unpublished and Other Prose Works by D.H. Lawrence*
Plays	*D.H. Lawrence: Three Plays*
Poems	*The Complete Poems of D.H. Lawrence*

R *The Rainbow*
SCAL *Studies in Classic American Literature*
SM *The Symbolic Meaning: The Uncollected Versions of "Studies in Classic American Literature"*
WL *Women in Love*
WP *The White Peacock*

D.H. Lawrence and Survival

Experience certainly does not justify us in limiting
the ruthlessness of individual "saviours of society" ...

T.H. Huxley, *Evolution and Ethics*

Introduction

The fiction of D.H. Lawrence, particularly the novels, brings to my mind an old English nursery rhyme about a "pretty little girl with the pretty little curl, right in the middle of her forr'id": when Lawrence is good, he is very, very good, but when he is bad, he is horrid. My initial reading of Lawrence's fiction occurred in a telescoped way: all of the novels and most of the other fiction in the space of about two years. What puzzled me most was how a novelist who could at times write so well was also capable of penning such fictional abominations as *The Plumed Serpent*.[1] This is not quite the same charge as that levelled by T.S. Eliot in *After Strange Gods* (and challenged by F.R. Leavis) that Lawrence "seems often to write very badly; but to be a writer who had to write often badly in order to write sometimes well" (qtd in Leavis, *D.H. Lawrence* 304). Nevertheless, one would be hard pressed to name any other acknowledged major writer with a spectrum of quality as broad as Lawrence's. I have yet to encounter a convincing conjecture that addresses this curious phenomenon – Eliot's class-related hypothesis is sheer snobbery, and his attempt to suppress *After Strange Gods* suggests that he eventually recognized that fact – and it is a question that I have come to believe is central to understanding the work of one of the twentieth century's most important but also most enigmatic writers. Perhaps the best-known book-length attempt to "explain" Lawrence's deterioration is Eliseo Vivas's *D.H. Lawrence: The Failure and the Triumph of Art* (1960). More recently, Wayne Templeton published an essay (originally a paper at a 1993 conference) that endeavours "to ascertain the reasons why

Lawrence's fiction changed so dramatically in manner … theme, and even methodology, and why, specifically, it changed during the war years" (184). His answer, that Lawrence's illness and what that illness meant for the writer's sense of self was the crucial factor, is an important contribution to understanding Lawrence's deterioration. Nevertheless, the exclusive focus on illness is too reductive, I think, while the essay's emphasis on the war years ignores the period immediately following the war, when, in my view, Lawrence made his most important attempt to change the direction of his fiction.

The key to Lawrence's uneven quality lies in the transitional works that led from the world of *Women in Love* (1920) to that of *Kangaroo* (1923). Of inconsistent quality themselves (they include some of Lawrence's best work, such as *The Fox*, as well as some poor efforts like *Aaron's Rod*), these works were in themselves a catalyst that transformed a writer of exploratory, experimental, and significant fiction into one who produced mediocre writing at best and, at worst, strident, shrill, preachy, and just plain poor work.[2] M. Elizabeth Sargent does well to point out that Lawrence's development as a writer toward leadership and misogyny looks much less linear now than it once did (177);[3] nevertheless, the decline in narrative power is still striking and the linear model convenient. Perhaps the most curious thing about the deterioration is that the changes that brought it about were ones that the author himself deliberately engineered. In the transitional works, Lawrence, in a sense, prepares us and himself for a shift from the setting of his early English novels to a different continent, one, moreover, that is about as far away from Europe as one can get. He also alters the focus of his work, and, even more importantly, he does so through a narrative strategy that allows him to revise his earlier works so that they accord with his new interests. Sometime in 1919, while waiting for permission to leave England and for the cloud of official suspicion that obscured his life during the war to lift, this most exploratory of British novelists began a housekeeping of his fiction that ended by about 1922 in a radical rebuilding. The renovation may have been necessary for Lawrence to continue as a writer and even to maintain his sanity, but in artistic terms it ended in something approaching disaster. The reconstruction is most obvious in the three novellas published together in 1923 (but written at the end of 1921). *The Fox*, *The Captain's Doll*, and *The Ladybird* are all based on earlier short stories that Lawrence returned to and, in lengthening into novella size, changed radically. There is even in *The Fox* a passage that can be taken as Lawrence's self-reflexive refutation of his own earlier, more optimistic self, the self that penned *The Rainbow* and that used the biblical and elemental connotations of the rainbow as a symbol of a better world attainable here on earth: "But the end of the rainbow is a bottomless gulf down which you can fall forever without arriving, and the blue distance is a void pit which can swallow you and all your efforts into its emptiness, and still be no emptier. You and all your efforts. So, the illusion of attainable happiness!" (*Fox* 69).

Symbolically speaking, there is now to be no more rainbow because the two elements that make up a rainbow, called fire and water in "The Two Principles" (1919) and associated, respectively, with male and female (*PII* 234), can no longer combine to create a radiant third thing. They are now seen to be in eternal conflict. The nightmare image of "a void pit" is natural enough for the son of a coal miner who never set foot in a mine, and the insatiability of the bottomless, swallowing pit is characteristic and central to Lawrence's writing at this time. That the passage above, on closer inspection, makes no logical sense – an empty pit that cannot be filled ought to be no *less* empty rather than "no emptier" – is emblematic of a confusion in the writing as Lawrence moved from marriage to leadership in his fiction. Similar imagery occurs in the roughly contemporaneous and extremely confused *Fantasia of the Unconscious*. Here, after an argument in favour of the submission of women to men in the usual terms of his denunciation of sex in the head, Lawrence suggests: "And if there *be* no further departure, no great way of belief on ahead, and if sex is the starting point and goal as well: then sex becomes like the bottomless pit, insatiable" (*FU* 194). In these passages and others like them, we may observe some of the covert terms that set the stage for the fiction of the leadership period: a feeling of confinement ("no further departure") and unappeasable hunger ("insatiable") linked with sexuality in an overall insistence on male purpose.

These terms are part of Lawrence's attempt, in the immediate post-war years, to imagine a revolution in the hierarchical organization of English society by making use of several doctrines from nineteenth-century evolutionary theory. Darwinism was the most significant, influential, and, of course, controversial new model of the world to arrive in Western culture since the sixteenth century, and it was absorbed into popular consciousness at the time that Lawrence was growing up. However, Lawrence's efforts to recast English social hierarchy in a convincing way ultimately fail due to his still-disturbing relationship to the earliest and most basic of all hierarchies: that between mother and child. The mother-child relationship from the point of view of the child is one of an overwhelming dependency centred upon the provision of food, nursing during illness, and the offering of an appropriate level of freedom. The signs of the failure are a confusion in the texts of the period and a telltale use of imagery of food, illness, and confinement; this imagery demonstrably undermines the fiction's conscious designs on the reader in terms of leadership or male purpose by constantly realigning the text toward the pre-oedipal concerns Lawrence seems unable either to resolve or to leave alone.[4] The result is a self-subversion that to some extent saves the fiction of the transitional period; however, the ideology worked out in these stories and novels lies at the centre of Lawrence's artistic fiasco as he moved from the transitional works to the leadership period of *Kangaroo* and *The Plumed Serpent*. The revision of the three novellas and the composition of the other fiction of the transitional period do not have an "underlying integrity of motive" as Albert J. Devlin claims the changes to *The Fox* have (82). The motive is confused, only partly conscious, and self-defeating. As a consequence, it is possible to find sites of contradiction in many of the fictional works of the transitional period and to use them to elucidate the underlying anxieties that ultimately cause Lawrence's narrative strategy to result in work of inferior quality.

Serious criticism of Lawrence's work can be conveniently dated from the publication of F.R. Leavis's *D.H. Lawrence: Novelist* in 1955. After several decades of New Critical readings of Lawrence's fiction following Leavis, including landmark studies such as H.M. Daleski's *The Forked Flame* and George H. Ford's *Double Measure* (both 1965), Lawrence criticism has continued in a multitude of directions, including, in recent years, studies from poststructuralist and postcolonial perspectives.[5] There have also been books and articles that attempt to "place" Lawrence's creative writing within his historical and cultural time. One example is Kim A. Herzinger's *D.H. Lawrence in His Time: 1908–1915* (1982), which attempts to understand Lawrence "inside the cultural matrix" of his era (18). Herzinger's book looks primarily at the Edwardian and Georgian cultural milieux and includes an analysis of the impact of futurism, imagism, and vorticism on Lawrence. Although the emphasis is, strictly speaking, on cultural and literary contexts in a delimited time period, it is nevertheless surprising that there is not one word in Herzinger's book on the continuing impact of Darwinism on the literary scene of the time and on Lawrence in particular. Such an absence is typical of studies on Lawrence, with a few notable exceptions that I will detail in chapter 1.

Unlike many of his contemporaries, who took Darwin's evolutionary model of life for granted, Lawrence engaged in a decades-long dispute with Darwinism only in the end to covertly adopt a number of evolutionary doctrines for his own fiction. Although there were elements in Darwin's theory that Lawrence never accepted, by the middle of his career he began to see his own individual works in terms of evolutionary process. His characteristic and exploratory setting up of contrasting ideas within his works to see which would survive the competition and the frenetic revision of individual works to allow them or force them to "evolve" into something "new" are examples of this development. Rather than make a clean break between the world of his marriage novels and that of his leadership works, he wished to forge an evolutionary change that required the transitional stage of writing represented by a number of works written roughly between 1919 and 1922, works that will be the focus of this study. I will argue that Lawrence's image of himself as a writer resulted in a narrative strategy in which his older concerns were made to evolve into his newer ones through the environment of fiction itself. Lawrence became his own "beast in the jungle," culling the weakest members of his conceptual herd in order to strengthen the whole and setting up characters who are sent through the alembic of a narratological survival-of-the-fittest test in order to distill the character traits he approved of. Eventually, we get the leadership fiction, whose peculiar mixture of utopianism and misanthropy corresponds to the optimistic and pessimistic readings (toward perfection and extinction, respectively) to which the idea of evolution is open, as Gillian Beer has pointed out (16, 22, 39*f et passim*).

Instead of the usual Lawrentian-charged words such as "blood," "spell," "darkness," and "leadership," which are trotted out in studies of the period, my analysis will make use of lesser-noticed terms such as "fitness," "confinement," "breeding," "eating," "illness," "garden," and, especially, "survival." Perhaps as much as anything else, these terms of reference and their Darwinian implications indicate the focus this study takes and how it differs from most analyses of Lawrence's work after the marriage novels. It would seem that one of the factors that has prevented us up to this point from seeing an important Darwinian influence in Lawrence's narrative strategy is a tendency to flinch from even the thought of the novelist as a kind of Josef Mengele of fiction. Bertrand Russell's well-known suggestion that one can draw a straight line from Lawrence to Auschwitz (Nehls i 283–4), an idea that may be more forceful than Russell intended, seems extreme and possibly personally motivated.[6] Darwin's legacy in great part is the concept of natural selection, an idea horribly perverted by the Nazis, most notoriously in their own selection process that, for example, divided prisoners between slave labour and murder at Auschwitz. Mengele at the railway ramp pointing people to the left and right is a searing image of our postwar age. We are thus reluctant to contemplate how a writer might (more or less) benignly resemble such a figure when a version of the principle of natural selection infiltrates the narrative strategy and the novelist sends his own characters to life and to death as Lawrence does in this period of writing. Yet, to understand what Lawrence was about in this, his most puzzling period of writing, one must explore just such a possibility. One earlier and disturbing piece of evidence for the concept of Lawrence as a selector for life and death can be found in a 1915 letter he wrote to Ottoline Morrell, in which he claims: "Sometimes I wish I could let go, and be really wicked – kill and murder – but kill chiefly. I do want to kill. But I want to select whom I shall kill. Then I shall enjoy it" (*L* ii 315). While few readers of Lawrence would take such a sentiment as anything more than an honest expression of a fantasy he recognized in himself during the distressing period of the war, it is nevertheless instructive to see in what directions his fantasies could take him at this time. Indeed, the most interesting affinities between Darwin and Lawrence appear to be those that Lawrence was largely unconscious of, so that even though Lawrence almost invariably dismissed evolutionary claims with vigour, it is because of rather than in spite of that dismissal that we do well to consider his own use of Darwinian concepts as a writer.

In selecting his characters for life and death, Lawrence was in large part working through a serious anxiety about his own survivability at an extremely vulnerable stage in his life. Despite the fact that he died at age forty-four, D.H. Lawrence was a survivor; he ought, by rights, to have died much earlier, even in childhood.[7] David Ellis paraphrases from "Getting On" Lawrence's belief that his mother "had decided that he was too weak to survive" (340).[8] His chronic illness and periodic acute health crises meant that personal survival preoccupied him to an unusual degree; certainly this is so if we go by the evidence of the creative writing rather than the letters, where he normally avoids any mention of his tuberculosis and refers instead to "bronchial" trouble. Lawrence's often-quoted dictum that "one sheds one's sicknesses in books – repeats and presents again one's emotions, to be master of them" (*L* ii 90) is most often used in a general way as evidence of the autobiographical nature of his writing, and rightly so. But it is possible to take the specific reference to sickness more literally than is usually the case. Apart from a restless wandering from country to country in apparent search of a climate beneficial to a consumptive, the only means Lawrence had at hand to try to master his own sickness was his writing. His dictum, then, might well be read as a proclamation that his writing is to be an arena in which the very things that threatened his survival would be opposed and defeated. In his final years' writing we can see him explicitly preparing for death in works such as "The Ship of Death" and *The Man Who Died*, but in an earlier period of his life, the time before the American sojourn, his doubts about his survival as an artist coincided with questions regarding his personal survival and the viability of his culture in such a way that we get fiction that is remarkable for its handling of the concept of survival. Furthermore, it is in the fiction of this immediate postwar period that we also see Lawrence using survival as a metafictional tool, part of a narrative strategy to reorient his own fiction in keeping with his shifting ideology and his largely unarticulated fears. Lawrence was more of a metafictional writer than he is generally given credit for, and in this transitional period especially metafiction became a significant resource for him for the simple reason that he was then seriously rethinking the form his creative work ought to take. As he composed creatively he was also thinking self-reflexively about the very enterprise upon which he was engaged and letting that thinking form part of the fictional text. In the posthumously published essay "The Novel," Lawrence even seems to equate the long fictional form itself with his own emaciated form and struggle to survive. The essay begins: "Somebody says the novel is doomed. Somebody else says it is the green bay tree getting greener. Everybody says something, so why shouldn't I! Mr. Santayana sees the modern novel expiring because it is getting so thin; which means, Mr. Santayana is bored" (*PII* 416). Later in the piece, Lawrence speaks of the necessity for the novel to be alive: "It can't exist without being 'quick.' The ordinary unquick novel, even if it be a best seller, disappears into absolute nothingness, the dead burying their dead with surprising speed." Tolstoi could not abide his own "really quick" characters and "loved to kill them off" (*PII* 420). Lawrence, on the other hand, will kill off his unquick characters in a fictional program of eugenics.

What happens, in brief, is that the conscious effort to make the old views "evolve" into the new, to jolt his art out of the cul-de-sac it has entered, sends Lawrence to Darwinian theory as the basis of his narrative strategy and in keeping with his ruthless outlook after the war. However, because of his own anxieties about survival, this manœuvre triggers a largely unconscious countervailing movement that is subliminally present in the imagery and odd twists in the plot, a movement wherein anxieties become projected onto the female characters in the fiction. It is this bi-level development with its attendant metafictional awareness that makes the works interesting. The stage is set for the next period of writing, however, in that Lawrence discovers that he must break through the resistance still remaining in the transitional works and opt for the unrestrained world of *Kangaroo* and *The Plumed Serpent*, where it is not the narrative strategy that is Darwinian but the entire world, as symbolized by the ritual of the bullfight at the beginning of *The Plumed Serpent*. The planned renewal of Lawrence's fiction ends up being reactionary.

I must emphasize that in focusing this study on some of Lawrence's more problematic works, and works, moreover, that clearly led to his worst fictional output, I am necessarily underplaying the best in Lawrence. Many of Lawrence's positions, such as his resistance to totalizing processes and his vehement opposition to the degradation of the natural world, still have much to tell us. I like and admire Lawrence's early and middle fiction. I agree with Barbara Schapiro that "[i]n his most interesting work, the narcissistic domination fantasies are tempered by the equally Lawrencean passion for mutuality and balance, for the otherness that limits narcissism and makes real relationships possible" and that "[i]n his greatest work, the subjective reality of the character ... competes with and ultimately withstands the polarized narcissistic fantasies that are projected onto them" (10, 17). Unfortunately, Lawrence was not always at his best. What I want to deal with here is precisely *not* Lawrence's most interesting or greatest work but work that is fascinating because it can help us begin to solve the puzzle of where and how this sometimes wonderful, sometimes enraging, but always achingly honest writer went so wrong.

In suggesting a Darwinian perspective to some of Lawrence's work, I certainly do not intend to reduce his fiction to that dimension alone nor to suggest that the criticism on Lawrence to date has been off base; indeed, although I do take issue at times with some previous interpretations of Lawrence's specific works, many of my positions build upon the insights of earlier critics, such as Judith Ruderman in terms of what she calls Lawrence's "unresolved pre-oedipal conflicts" (8) and Margot Norris and Roger Ebbatson with respect to Darwinian influences on Lawrence, to name three. What I would like to do in the following pages is tease out and expand upon a neglected aspect of Lawrence's writing within a critical world where our understanding of Lawrence is multi-dimensional. The first chapter reviews what little work has been done on Lawrence and Darwin and begins the construction of my own argument in terms of how Lawrence used Darwinian concepts in a particular period. The next three chapters look at fiction where Lawrence clearly constructs a narrative strategy under the impress of Darwinian doctrine in order to build a bridge between the achievement of his marriage period and the place that he feels he must now journey to in the postwar world. His fiction and his metafictional comments in the three *Ladybird* novellas,[9] in the novels *The Lost Girl* and *Aaron's Rod,* and in the short stories of the *England, My England* volume reveal a Lawrence preoccupied with survival and writing against his own best instincts in a defensive enterprise that ultimately leads to his nadir as an author.

The nine works of fiction I discuss in detail are not presented in chronological order. Chronology can be rather tricky with Lawrence, who never saw a typescript or manuscript of one of his works without developing an itch to revise it. Given Lawrence's frequent difficulties with publishers concerned with the policies of circulating libraries and the tastes of censors, publication dates are an extremely unreliable guide to where a work belongs in Lawrence's development as a writer. Even if we were to take completion dates, as far as they are known, we would be faced with the often impossible task of sorting out one portion of a revised novel or story left unchanged from an earlier version from a portion that has been transformed radically. Consequently, I consider the transitional period of writing as one unit for the purposes of addressing the broader issue of Lawrence's use of Darwinian concepts in the project of moving his fiction from one world to another. The order of presentation is based on the practical consideration of the most effective sequence to illustrate my argument.

1 Lawrence and Darwin

In *Darwin's Plots* (1983), Gillian Beer examines Darwin's rhetoric and how his mode of expressing his discoveries had inevitable consequences for the fiction that followed the publication of his major findings. Because of its preoccupation with time, as Beer points out, evolutionary theory has "inherent affinities with the problems and processes of narrative" (7). It affected the novel of the nineteenth century "not only at the level of theme but at the level of organisation" (8). In particular, "[t]he *methods* of scientists become the methods of emplotment, and scientific theories suggest new organisations for fiction" (161). George Levine, also focusing on the nineteenth century, puts the issue another way: "the psychology based on evolutionary biology introduces elements of discontinuity that in turn led to revised notions of the self, and consequently of character in fiction, and of course of the way narratives move and are resolved" (7).[1] Beer proceeds no further than Thomas Hardy in her study, but it is safe to say that the advent of a new century does not wipe out the influences prevalent in the preceding years. Thomas Hardy was saturated with evolutionary concepts, but so were his immediate successors in the English novel.

H.G. Wells's *Ann Veronica* (1909), for example, shows the effects of the application of Darwinism to eugenics, the eugenics movement in England being at its height at the time (Trodd 63). Edmund Gosse's autobiographical *Father and Son* (1907) is a virtual primer in the conflict between religious faith and Darwinian theory in the person of Gosse's father, Philip, a leading member of the Plymouth Brethren and an outstanding marine biologist. A few years later, in *Howards End* (1910), E.M. Forster could assume a ring of authenticity in the ears of his public for the dialogue in chapter 22 where Henry Wilcox and Helen Schlegel debate the question of social equality in terms made possible by an optimistic reading of Darwin (not the only possible reading, as Hardy's deep pessimism suggests). When Henry has retreated, Helen complains to her sister, Margaret, that "'he must be one of those men who have reconciled science with religion … I don't like those men. They are scientific themselves, and talk of survival of the fittest, and cut down the salaries of their clerks, and stunt the independence of all who may menace their comfort'" (193).

Oddly, D.H. Lawrence, the writer most identifiable as Hardy's heir in the English novel, has generally been viewed as immune to Darwin's influence. Gosse's book came out at about the time that Lawrence was taking his first, tentative steps as a published writer.[2] However, although Forster's Henry Wilcox is, in a number of ways, a forerunner of Lawrence's Gerald Crich, there is nothing like this overt discussion of Darwinism in *Women in Love* or any other novel by Lawrence. Few have asked why this should be so. Roger Ebbatson has pointed out with respect to Lawrence that Darwin's theories "would generate an imaginative response even where (or especially where) the novelist felt impelled to an act of rejection" (*Evolutionary Self* xxi). But why has Lawrence's anomalous failure to engage with Darwinism in his fiction not piqued much critical interest?

One explanation for the historical reluctance to link Darwin and Lawrence, even negatively, is the many hostile comments Lawrence allowed himself in his battle with Darwinism over the years. We know that Darwin was one of Lawrence's early adult interests (Worthen, *The Early Years* 171). Having carefully read Darwin, Haeckel, Huxley, and Spencer in his youth and for a time been an enthusiast, Lawrence later reacted against the evolutionists, claiming time after time in his essays, letters, and travel writing that the tenets of their theories were all wrong. Aldous Huxley reported that science in general and evolution in particular was "a matter in which [Lawrence] absolutely refused to take a rational interest" (xv). What Lawrence seems to object to most in evolutionary theory is the notion that, in the words of his poem "The Rose of All the World," "[t]he seed is purpose, the blossom accident," that is, the theory's postulate of the insignificance of the individual in the chromosomal scheme of things (*Poems* 218). Perhaps his most explicit repudiation of Darwinism comes in *Mornings in Mexico*, where he writes: "Myself, I don't believe in evolution, like a string hooked on to a First Cause … I prefer to believe in what the Aztecs called Suns: that is, Worlds successively created and destroyed" (*MM* 12). Occasionally, however, he implicitly endorses a Darwinian reading of nature, such as when, toward the end of *Women in Love*, Birkin engages in the well-known speculation about the "creative mystery" and how it gives rise to finer forms to replace those that fail to develop creatively (*WL* 478–9). The creative mystery Birkin alludes to here may be some sort of jealous Jehovah who dispenses with "sinful" species and individuals, but the "sin" is clearly the failure to evolve.[3] Moreover, as George J. Zytaruk has noted, Lawrence's "account of the origins of life" and "view of individuality," at least in the Hardy essay, "are consistent with the theory of evolution" (244).

The disinclination to associate Lawrence with Darwin is also likely due to the fact that, although fully one-third of D.H. Lawrence's short life was lived with Queen Victoria on the throne, we do not generally view Lawrence as a transitional figure in the same way as Hardy is so obviously one. We tend to think of his impact on such matters as censorship as a comparatively recent development because society's "official" acceptance of Lawrence's explorations in the realm of human sexuality did not occur until some thirty years after his death, that is, in 1960. Lawrence's roots, however, go back to the time immediately following Darwin's publication of *The Origin of Species*, a full century earlier than the *Lady Chatterley* trial but only a single generation before Lawrence's birth. Emile Delavenay exaggerates perhaps when he suggests that Lawrence's psychology is closer to the nineteenth than to the twentieth century (*Edwardian Transition* 4), but one sympathizes with his attempts to correct a widespread misconception.

Darwinism is a materialist theory that might have attracted the Lawrence who so emphasizes the body and its needs. It is developmental and capable of Utopian interpretation. To an anti-idealist romantic like Lawrence it should have held great appeal. As a young man, from what we can tell, Lawrence did accept the basis of Darwin's theory, the concept of natural selection. In "Art and the Individual," originally a paper delivered in 1908 to the Eastwood Debating Society (Worthen, *The Early Years* 178), Lawrence speaks of the whiteness and wings of the swan as expressive "of the great purpose which leads the swan to raise itself as far as possible to attract a mate, the mate choosing the finest male that the species may be reproduced in its most advantageous form" (*PII* 224). This emphasis on the relatively benign aspects of Darwin's theory was not to last, however.

One of Lawrence's objections to Darwinism, as Ebbatson has suggested ("Spark" 90), has to do with what might be seen as a determinism constraining individual freedom and an emphasis on the collective rather than the individual. That a theory of "recapitulation" (the phases in evolutionary process repeated in individual development) arose from Darwin's ideas does point to the inherent determinism of the theory. A scene in *Aaron's Rod* dramatizes another aspect of Lawrence's quarrel with Darwin. Aaron is discussing with his host in Novara, Italy, Sir William Franks, his own and Rawdon Lilly's views of life. Aaron's seems most to accord with a Darwinian view of things: "'But I believe in chance,'" he declares. If his luck runs out, he continues, then he may simply die (*AR* 143). Sir William disagrees that chance is the operating force in the universe, but admits that at least Aaron's view has the virtue of logic to it lacking in Lilly's. Lilly seems to believe in a benevolent providence that rewards an individual's fidelity to his or her own vital spirit. Such a notion is close to Lawrence's own (and traceable, arguably, to the romantic privileging of authenticity) and may suggest a key point of disagreement with Darwin. The deck is stacked here in that Sir William, a firm opponent of Lilly's view, is shown to be near death, as if he were being punished for denying his own vital self, thus confirming Lilly. Such a narrative manoeuvre of portraying winners and losers is typical of the transitional period, as we will see.

The few critics who do consider Lawrence and Darwinism tend to overlook one further central objection the novelist had: the theory's emphasis on the dimension of time. Although in *The Origin of Species* Darwin often discusses territory and the changes in environmental pressures on a species depending upon where that species' natural territory lies, the basic conceptual framework of the theory and what made it so controversial in its day is not space, of course, but the portrayal of the effects of immense intervals of time on living species. The central assumption of Darwin's theory is that chance mutations that happen to provide a slight edge in the lottery of survival will, given enough time, create fundamental changes and, ultimately, new species. It is the same temporal ordering of reality that Darwin takes as his given that, according to Beer, makes evolutionary theory so relevant to the development of literature, manifestly a temporal art. The *Origin* is a narrative, "because what it describes cannot be correctly described except through the medium of time" (Beer 65). Evolution's breaking of temporal (as well as species) boundaries in its view of time as one continuum might seem attractive to the author of "The Border-Line," a man who spent his own life crossing frontiers and breaking taboos in his art. But the fact is that to so impatient a man as Lawrence, so keen on the rapid betterment of humankind, the Darwinian concept that such a development can occur only over vast stretches of time and, in modern terms, only through the passing on of a gradually amended gene pool through reproduction would have been most uncongenial.[4] Lawrence saw an overemphasis on time to the detriment of space as one more example of his culture's loss of balance. For Lawrence, Hardy's Angel Clare resembles nineteenth-century evolutionists in his conception of the world in monistic terms, in the terms of "one spirit or principle starting at the far end of time, and lonelily traversing Time. But there is not one principle, there are two, travelling always to meet, each step of each one lessening the distance between the two of them. And Space, which so frightened Herbert Spencer, is as a bride to us" (*P* 485). The language Lawrence uses here in the Hardy study in describing the incremental closing in toward each other of the two principles and in his characterization of space as a bride brings to mind the highly charged and effective scene in the fourth chapter of *The Rainbow*, "Girlhood of Anna Brangwen," where the two lovers, Anna and Will, in putting up some oat sheaves, move closer and closer to each other until they meet in an embrace in what amounts to a ritualized courtship dance as well as something of a battle of wills (*R* 113–16). The scene dramatizes the spatial turn of Lawrence's imagination even in a novel so concerned with the generational passing of time and, in the present context, suggests why Lawrence objected to aspects of evolutionary theory. He internalized the Darwinian conception of change, rendering it spatially contingent (that is, requiring the right *person*, not the right amount of time) just as he eventually located it in the structure of his fiction rather than in the genes of his characters.

That Lawrence was actively engaged with the idea of evolution, something one would expect of any intellectual born in the late nineteenth century, has also been obscured by the embeddedness of Darwin's influence on Lawrence. For example, Lawrence's frequent use of gamekeepers has often been noted but seldom associated with Darwin. Worthen remarks that Lawrence would have seen gamekeepers in the area around Eastwood and that such figures "preserved the game of the landowners against the predations of the industrial class who lived in the towns and villages" and that the keepers themselves belonged to the working class (*The Early Years* 227). The focus on class is well taken and is usual in discussion of Lawrence and gamekeepers. The gamekeeper, indeed, acts as an agent of control for a certain privileged class of people. Worthen's contribution to this minor issue is in indicating the difficult class position and divided loyalties many keepers must have felt. Nevertheless, he, like most other commentators on Lawrence's gamekeepers, glides too lightly over an obvious but neglected point: that the keeper also interferes in the natural balance of survival among the animals, and even in the human struggle for survival. A gamekeeper, after all, whatever he may do in his leisure time (and Lawrence certainly had a lot to say about that in *Lady*

Chatterley's Lover), is employed to keep game, an important source of food, to cull it for his master's use, and to keep it away from anyone else who might like to exploit it. In *The White Peacock*, Annable, Lawrence's first gamekeeper figure, is ferocious in his defence of his master's rabbits: "Determined poaching was carried on in the Squire's woods and warrens. Annable defended his game heroically. One man was at home with a leg supposed to be wounded by a fall on the slippery roads – but really, by a man-trap in the woods" (*WP* 125). Part of Lawrence's fascination with the figure is undoubtedly connected to the gamekeeper's power over life and death in the animal world; he would certainly have been alert to the gamekeeper's role in the struggle for survival. In his essay "Reflections on the Death of a Porcupine," he writes: "Wherever man establishes himself, upon the earth, he has to fight for his place, against the lower orders of life. Food, the basis of existence, has to be fought for even by the most idyllic of farmers … Food, food, how strangely it relates man with the animal and vegetable world! How important it is! And how fierce is the fight that goes on around it" (*PII* 465). Despite the advantages possessed by the British landed classes, including the employment of gamekeepers, it is not class privilege that, for Lawrence, ultimately determines the winners and losers in the processes of natural and sexual selection. In *Lady Chatterley's Lover*, famously, it is the gamekeeper rather than his master who wins the battle of sexual selection. In the "Porcupine" essay, it is a short jump for Lawrence from the idea of a hierarchy of existence among various species – the "higher" and "lower" orders of life – to such a ranking within the human species. And he takes the plunge, for example, when he discusses what he calls "the cycles of existence," in which the central question is "*Can thy neighbour finally overcome thee*?" If my neighbour prevails, "then he belongs to a higher cycle of existence. This is the truth behind the survival of the fittest" (*PII* 468).

Darwin's embedded influence on Lawrence is perhaps most significant in the area of narrative strategy. What Anne Fernihough says regarding Lawrence's use of Nietzsche and Roger Fry applies equally to his handling of Darwin: "Lawrence often stages overt disagreements with philosophers and critics only to agree with them at a more covert level" (147). In fact, evolutionary ideas affected not so much Lawrence's thinking or that of his characters in any obvious way but primarily and less obtrusively the narrative structure of his fiction, particularly, I would argue, in the fiction of his 1919–22 transition to the leadership works. In *The Fox*, Henry Grenfel may be said to be a variant of the gamekeeper figure. He intervenes in the struggle between fox and hens for the benefit of human industry but then proceeds to become a kind of predator himself in the human world. In other fiction, the gamekeeper role becomes even more internalized within the textual structure as the focus moves from the world of nature in the grip of the battle for survival to the social realm, which, especially after the war, in Lawrence's eyes has become a place where only the fit survive.

Whatever real areas of agreement or disagreement there might exist between Darwinism and Lawrence's thinking, he finally had to denounce the nineteenth-century doctrine for the simple reason that it ultimately proved too dangerous to endorse it. What turned Lawrence away from Darwinism in the end, I think, was the frightening idea that, by Darwinian standards of natural selection, Lawrence himself was expendable. He had to deny the entire theory in order to defend against the implications of his own lack of robustness, his "unfitness." However, when we come to the transitional period of Lawrence's writing, what has been a successful campaign of defensive denial breaks down under the pressure of the compound crisis of Lawrence's own deteriorating health, the problematic state of his marriage, and his loss of audience in Britain. The result was an extreme reaction whereby Darwinian theory was given a twist and embraced in the project of survival. Lawrence, then, felt threatened by Darwinism because, according to its tenets (which he knew very well), he was far from a paragon of fitness for survival. But just as an abused child will unconsciously identify with the aggressor as a coping mechanism so Lawrence accepted aspects of Darwinian doctrine and applied them in and to his own writing. This ambivalent attitude is similar to his view of the war and stems from essentially the same psychological defence: the war was an anathema most of the time, but occasionally it was welcomed as the force that would cleanse Europe of its disease.[5] Lawrence himself would be made stronger emotionally if he purged those weaker, "female" traits of his early work that led one reviewer of *The White Peacock* to indentify the gender-neutral initials "D.H." as likely belonging to a woman, as Lawrence wryly noted in a letter to his sister Ada (*L* i 229). His work would become strong enough to withstand the onslaught of moronic censors and hostile critics if it were to undergo a "purifying" process of change.

Lawrence's ambivalence toward Darwinism had some manœuvring room in part because of a semantic peculiarity. In the eighteenth century, the word "evolution" meant ontological development, that is, the changes undergone by an individual in a lifetime. Only after 1830 did the term begin to refer to species development, or phylogeny. As a result, "[t]he blurring of the distinction between ontogeny … and phylogeny … in the single term 'evolution' proved to be one of the most fruitful disturbances of meaning in the literature of the ensuing hundred years" (Beer 15), a hundred years, in fact, that ended in the year of Lawrence's death. Lawrence, in effect, takes advantage of just this philological instability in the word "evolution" to internalize phylogeny, as it were, and make it refer to the individual. The individual for Darwin and other evolutionists is the means for the further development of the species and in himself or herself is a dead end. As Beer puts it, "[t]he individual organism does not evolve in the course of its life. Though it takes part in the evolutionary process, it does so only through generations not through any happening within its own life cycle" (43). How does Lawrence come to terms with such a view? Ebbatson suggests that Lawrence transforms evolutionary thought itself into a means to an end, the means of generating "the Lawrencean novel" (*Evolutionary Self* 76). His extension of rebirth to the individual from the species in effect inverts Darwin's transference of *extinction* from the individual to the species. We might view Lawrence's personal symbol, the phoenix, as emblematic of his desire for evolutionary change within the individual and of his wish to believe in the possibility of immortality. This is so at least until the point where his bitterness and impatience lead him to go beyond the exploration of individual "allotropic" development of character to an envisioning of brutal socio-political solutions.[6] By the leadership works, atomistic development is no longer sufficient for the simple reason that the containing culture oppresses the individual; the entire social hierarchy must be overthrown.

In the transitional period, there is a reassessment of hierarchies of all sorts as Lawrence attempts to come to terms with what he believes to be the malaise of his time in European civilization. Although he never entirely abandons his exploration of the relationship between men and women, his natural home as a writer of fiction, the shift of focus as he moves toward the concerns of *Kangaroo* and *The Plumed Serpent* is, at least on the surface, in the direction of social organization. Despite a consistent emphasis on the individual, Lawrence often tried to imagine a collective betterment and the means by which that might come about. For without a supportive society it is difficult for the individual to gain fulfillment (as Lydia Lawrence knew all too well, isolated in her imagined superiority from the other collier wives of Eastwood). From *Women in Love* on, heterosexual love relationships become, in effect, the meeting ground between the individual and the collective, and an approved couple can be imagined as the core of a new society, as Lawrence tries to kickstart his Utopian community.

In turn, this development allows Lawrence to revisit more easily a theory that tended to downplay the importance of the individual, that tended to place "individualism … under a new and almost intolerable tension by [its] emphasis on variability" (Beer 127). What Darwinian transformation allows Lawrence eventually to do in the portrayal of society is to imagine "saved" and "damned" groups destined for survival and extinction respectively.[7] In this sense and others, the revolution he seeks is truly a biological one, as Mensch has suggested (162),[8] although we must remember in fairness to Lawrence that he did not have the benefit (or burden) of our historical knowledge, a knowledge that makes the genocidal implications of his fantasies difficult to avoid. Darwinian theory allows Lawrence to evolve his old vision into his new by selective culling until the latter has little recognizable relation to the former. Certainly, it is very difficult to see how the same writer could create both *The Rainbow* and *The Plumed Serpent*.

Lawrence was always interested in class, and he lived at a time when the British class system was being destabilized, in part by the widespread propagation of Darwin's ideas in the *Origin*. That the class structure was based on such flimsy foundations as the outmoded belief in the inheritability of merit made the structure all the more vulnerable. Humanity's lowly beginnings among the primates had implications for the conception of class that were uncomfortably obvious to many Victorians (Beer 63). Despite his occasionally obsequious attitude to various members of the English upper classes (especially when he needed money or a place to live), Lawrence's abhorrence of the class system of English society is undeniable. It was that system that he later in life, rightly or wrongly, saw as largely responsible for his parents' bitter quarrels. In *The Captain's Doll*, a working-class driver taking middle-class tourists to a glacier in the Tyrol Mountains shows human warmth amid a world of ice only when he makes a stop at his own home. This prompts the narrator to exclaim: "Deep, deep is class-hatred, and it begins to swallow all human feeling in its abyss" (*Fox* 128).[9] Lawrence's imaginative efforts to rethink the very concept of class were far more radical, I would submit, than even those of contemporary socialists (or supposed socialists) such as H.G. Wells, who in *Tono-Bungay* (1909) expresses the conventional view – admittedly through the mouthpiece of his narrator, George Ponderevo: "Nothing can make an aristocrat but pride, knowledge, training, and the sword" (62). Lawrence, on the other hand, in an August 1915 letter to Lady Cynthia Asquith writes: "there are aristocrats and plebeians, born, not made. Some amongst us are born fit to govern, and some are born only fit to be governed …

But it is not a question of tradition or heritage. It is a question of the incontrovertible soul" (*L* ii 379). Wells, however, was much more practical in his views of class and the possibilities of bringing about meaningful change than Lawrence ever could be. Lawrence will use Darwinian concepts in his fiction to undermine the hierarchy of class, but then proceed to attempt to put in its place a new "class" system based on what he conceived of as the leadership of a "natural" aristocracy. "[I]t must be a system of culminating aristocracy, society tapering like a pyramid to the supreme leader," he declares in *Fantasia* (*FU* 182). Even in what is primarily a domestic drama in the story "Hadrian," the name of the titular protagonist clearly alludes to leadership, although it is difficult to see any such potential in the manipulative and unlikable young man in that story.

In reconceiving social hierarchy, Lawrence also makes occasional fictional gestures in the direction of what Margot Norris calls the "biocentric tradition." According to Beer, Darwin's enterprise has the effect of displacing or decentring humanity from its own conception of the reality of the physical world, an anti-anthropocentrism that undermines the very concept of hierarchy (60). This is the niche in Darwin that Norris exploits in her *Beasts of the Modern Imagination*. Norris explains that the biocentric tradition, an interrogation of all anthropocentric assumptions, included Nietzsche, Kafka, Ernst, and Lawrence and barely survived the thirties in art because of its paradoxical anti-art stance (1–3). It reverses the hierarchical positioning of human and animal; it treats the biological world as the real; and it engages in "animal gestures" in art. The founder of the tradition was Charles Darwin, who replaced the concept of Nature as a machine with his theory of natural selection, one that removes design and intelligence as the source of life and replaces them with organic forces working by chance over time. Darwin, in short, liberated biology from physics (Norris 6). One tends to think of Darwin as wholly an empiricist, perhaps because his contemporary opponents were so obviously anti-materialist. However, as he himself recognized, many of his postulates were counter-intuitive, and, as Norris points out, "his facts are not so much observable phenomena as relations between forces" (37).

One may infer from what Norris says that Lawrence's fight against the excesses of the mechanical principle and Darwin's revolutionary rejection of a mechanical view of nature in favour of an organic one were more similar than they first appeared: "And, in the very bloom of the Industrial Revolution, the machine and the rational intelligence that designs it become philosophically inimical to the burgeoning vitality of Nature and of living things" (Norris 7). Darwin's "ontological response to the wilderness" (35), moreover, is close to

Lawrence's reaction to the "spirit of place." That social Darwinism could later be used to support the very capitalist enterprise whose world view Darwin's theory undermines (by its implicit challenge to the hierarchy that sanctions human exploitation of the natural world) is one of the ironies of history and one that also helps to explain Lawrence's apparent rejection of Darwinism. Their territory as writers is virtually the same: "the difference between human and animal, male and female, Nature and culture" (Norris 37). Gerald Crich in *Women in Love* and Jack Callcott in *Kangaroo*, to name only two Lawrence characters, see the world as a jungle in which "the stakes are survival" (Mensch 200).

One of the merits of Norris's thesis for an understanding of Lawrence is that it places his propensity to support or at least to imagine a positive outcome to violence in his later writing within a comprehensible framework. For biocentric thinkers, claims Norris, animal violence is not to be tamed but "restored to its amoral Dionysian innocence" (10). The romantic concept of organic form in art (which Lawrence in large part inherited from Coleridge)[10] is given a biological sanction in Darwin's observation that natural forces produce living forms: "Specifically, it is natural selection and sexual selection that give the elephant her trunk and the stag his horns" (15).[11] Furthermore, as we can see to some extent in Lawrence, biocentric art "tends to flourish in the minor works of major authors," in such forms as short stories and novellas (20).

Norris sees Lawrence's writing (like Max Ernst's art) as fundamentally biocentric and thus Darwinian. His practice

> is to insert aspects of Darwinian living form, such as temporality, mutability, and susceptibility to chance, into literary conventions, thereby transforming character into multiplicitous and fluid 'allotropes,' eschewing novelistic closure, propelling events by irrational mechanisms of impulse, nostalgia, and free association that Lawrence often cloaks in the language of magic or spells. (18)

One might even argue that Darwinism made possible the kind of character transformation that distinguishes Lawrence from his immediate predecessors in the English novel. Even as late as George Eliot, the more or less static character derived from the eighteenth-century novel (and more remotely the comedy of manners on the stage) is pretty well the norm. The small, incremental changes that occur in the character of Adam Bede, for instance, or even Dorothea Brooke, are fundamentally different from the apocalyptic transformations that beset even so relatively unambitious a Lawrence character as *The Rainbow*'s Tom Brangwen (leaving aside the question of which character and which changes seem more "true to life"). As Delavenay has written, the original idea of "The Sisters" (the work-in-progress that eventually became both *The Rainbow* and *Women in Love*) was to show the development of the individual to a "higher evolutive state in which there is a partial abandonment of self" (*Edwardian Transition* 60). A George Eliot character may show the growth of a flower from bud to blossom, so to speak, where the end is inherent in the beginning; a Lawrence character will often feature an unpredictable transformation from tulip to rose. The Lawrentian evolution of character can sometimes seem so abrupt – as with Mabel Pervin and Dr Fergusson in "The Horse-Dealer's Daughter," for example – that it stretches verisimilitude to the point of snapping. What is "modern" in Lawrence after *Sons and Lovers* is not only (as he claims in his June 1914 letter to Edward Garnett) his liberation from the moral scheme of fiction that constrains Eliot and even Hardy but also the manner in which his characters strive to fulfil their full potential (including most emphatically in the sexual realm). Lawrence's characters – at least, his protagonists in the novels – independent of the normative cues that the moral scheme traditionally imposes in the English novel, will sometimes go through successive stages, the "allotropic states," so basic that the best analogy is Darwin's explanation of how new species come about through a long process of greater and greater change until there is little to suggest the parent form in the descendent. At times such changes are accelerated from the sedate speed contemplated by evolutionary theory to the point of appearing from one generation to the next as in *The Rainbow,* and then they are the source of much intergenerational conflict and bitterness.[12] Because Lawrence is impatient to see change come about, he is intolerant of his own characters who will not evolve. As Ebbatson writes, the Lawrence characters who freeze in their development (as Gerald Crich freezes literally) "are abandoned by their creator" (*Evolutionary Self* xiv). Though such a propensity may suggest a Victor Frankenstein-like act of desertion, this narrative strategy is influenced not so much by the eighteenth-century idea of the author as Creator as by the notion, derived from the nineteenth-century work of Darwin, of the author as a force of natural selection.

Norris argues that Darwin's reading of nature takes on the quality of a palimpsest, which is "written by force and desire, by the need to preserve some words and destroy others, by the wish to remember selectively and to forget" (41). Such a formulation is highly suggestive when considering Lawrence as a metafictional writer. For, beyond his use of Darwinian concepts in his treatment of individual characters and the conception of fictional character in a more abstract sense,

Lawrence self-reflexively applies to his own writing this same method Norris describes in Darwin, allowing some ideas and characters to survive the crucible of conflict, while others perish. Lawrence himself becomes, in effect, the animal predator, or perhaps more accurately the breeder or cultivator whom Darwin speaks of as practising a form of human selection that is akin to the natural kind (47–57). The writer's "animal gestures" are, then, part of the very technique of his art, a determining aspect of the very structure of his fiction, and it is surely no mere coincidence that, during the period of writing in question, many of Lawrence's works of fiction display totemic animals (the fox is perhaps the most obvious example) who become actual or symbolic agents of change within the story and, arguably, metafictional embodiments of the "animal gestures" Norris is talking about.[13] Lawrence's self-reflexivity may not be as overt as the Ted Hughes poem "The Thought-Fox," for example, but there can be little question that Hughes, an admirer of Lawrence, wrote his poem with Lawrence's metafictional predilections in mind.

The trouble is, as Norris points out with reference especially to Lawrence's essays, there is an inherent irony or even paradox in the biocentric artistic activity that is clearly evident in Lawrence's polemical writing, where he uses rationality to launch an assault on reason, deliberate argumentation to sponsor spontaneity. He has similar problems in his creative writing of the time: "How can Lawrence argue, in the conventional rhetoric of argumentation, that being is not subject to argumentation? Lawrence's problem, like Nietzsche's, is to make his writing constitute an act, a verb rather than a noun, a gesture rather than an object" (Norris 173). Although, in the view of many of Lawrence's contemporaries, he succeeded all too well in making his writing an act – an obscene one they did not care to contemplate – the problem of contradiction remained acute for Lawrence himself and poses one still for the critic of his writing. Norris seems to endorse Lawrence's non-rational way of argumentation, for example in his essay "The Novel," where he distinguishes between the "quick" and the "dead" without giving a logical explanation because he wants to avoid a positivistic description of life (Norris 178). With his leadership fiction in mind, however, it is difficult to ignore the sadistic overtones in the metaphor of a table as "a mere disconnected lump, like a cut-off finger" in that essay (*PII* 420),[14] not to mention the disturbing division into living and dead, which, when seen through the lens of a Darwinian narrative strategy, resembles fictional victims and victimizers.

Where Norris situates Lawrence within a Darwinian tradition that displaces humanity from its anthropocentric position, Roger Ebbatson, who has done more work on the connection between Lawrence and evolution than anyone else, puts Lawrence within the "Nature Tradition" in the development of the English novel. While this may sound similar to Norris's biocentric tradition, in fact, its central concern is not the place of nature itself in human conception nor a vitalist concern but the reestablishment of humanity's central position within nature – in other words (from Norris's perspective) very much a recuperated anthropocentric one. Darwin's thought is crucial here since the "ground-plan" for this tradition is "the dialectic between Romantic and Darwinian evaluations of Nature" (*Nature Tradition* 20). In *Lawrence and the Nature Tradition* (1980), Ebbatson locates the "Nature Tradition" in the English novel between the publication of Darwin's *Origin* in 1859 and the beginning of the First World War in 1914. It draws upon "a complex body of thought and feeling about Nature" derived in part from the Romantic conception prevalent in Darwin's own day, New England Transcendentalism (especially Emerson), the Victorian notion of the open road, and "the debate about evolution" (4). The result is that, after Dickens, English novelists are heirs to both a Romantic and a scientific view of nature, and the relationship between the human and natural worlds takes pride of place in novelists from Meredith to Lawrence. These novelists attempted a synthesis "of Romantic visions of Nature as revelation and deliverance and Darwinian pictures of man's ordeal in the natural world." The Nature Tradition in the English novel, in Ebbatson's view, culminates in Lawrence (25, 26). Briefly stated, Ebbatson argues that Lawrence's reaction to Darwinism is threefold: a frequent depiction in the fiction of a struggle involving survival and extinction in the organic world, the dramatization of a Huxley-inspired sexual rivalry, and the revision of the concept of the interrelatedness of all of Nature into the Lawrentian concept of human relationship (33–4).

In *The Evolutionary Self: Hardy, Forster, Lawrence* (1982), Ebbatson extends his argument from the 1980 work on several fronts. Here he argues that Darwin's theory becomes a kind of "mental habit" for writers like Hardy and Lawrence that allows them to develop a new way of rendering temporal experience (x). Such studies as E.B. Tylor's *Primitive Culture* (1871) and Max Nordau's *Degeneration* (1895), written in the wake of Darwin's pioneering work, look at the dark underside of the implications of Darwinism and suggest, for example, the tragic potential in the failure to adapt. As the idea of fixed species gives way to some sort of concept of continual process, essence is replaced by becoming, and the novel takes on renewed importance as the analysis of the individual in society (xi-xiii). Key evolutionary terms such as struggle, fitness, and survival become translated in the novel into "a kind of myth of origins" (xiv). Of primary significance for the present study of Lawrence is Ebbatson's observation that the evolutionary perspective may have allowed certain novelists "to see themselves as the result of their own earlier work" (xvii), in other words to place themselves *as writers* within an evolutionary framework. Ebbatson presumably would not buy Norris's argument that Lawrence is engaged in much the same project as Nietzsche. In fact, the Nietzschean representative of a master race, Gerald Crich, who justifies his methods along social Darwinian lines, "embodies Lawrence's critique of a civilisation built out of repression and functioning through survival of the fittest" (99).

Most recently, in "A Spark Beneath the Wheel: Lawrence and Evolutionary Thought" from 1987, Ebbatson suggests that while Lawrence was deeply affected by evolution, he "simultaneously sought to redress its causal determinism with a more romantic apprehension of nature" (90). Lawrence's men and women "interact in a quasi-Darwinian way" (91). Huxley's concept of the natural world in continual ferment gets "translated, in Lawrence's theory of fiction, into a breakdown of traditional character typology" (91). Lawrence derives from Herbert Spencer his concept of the process of individuation and the pattern of "evolution and dissolution" so admirably captured in *The Rainbow* and *Women in Love* (93–4). Although Lawrence took exception to Ernst Haeckel's materialism and his mechanical view of the universe, he derived from him his early concept of the relationship between the individual and the community, his sun worship, his quarrel with Christianity, and his view of the importance of sexuality (96–7). In short, Ebbatson declares that Lawrence, all of his disclaimers notwithstanding, "was affected by the theory [of evolution] at the deepest imaginative level" (100). Indeed, Lawrence's "peculiar hallmark" as a writer is "the reconciliation of evolutionary theory and nature mysticism" (102).

Ebbatson's research is useful in a number of ways. It may serve as a corrective to Norris and vice versa. His observation that, for Lawrence, evolution "cuts both ways: it may be felt as loss of instinctive being or as hope of renewed life" (100) accords well with Beer's description of the coexisting optimistic and pessimistic potential in the theory of evolution and further serves to suggest how a Darwinian *method* of narration will be attractive to a writer who wishes to imagine a world in which the fit and the unfit, according to his own definition, will meet their merited fates. He recognizes that evolutionary principle shapes *The Rainbow,* although he does not extend the observation (102). Ebbatson quotes the same passage from

Lawrence's "Reflections on the Death of a Porcupine" that he cites in *Lawrence and the Nature Tradition* as an example of Lawrence's rewriting of Huxley: "The primary way, in our existence, to get vitality, is to absorb it from living creatures lower than ourselves ... There are many ways of absorbing: devouring food is one way, love is often another. The best way is a pure relationship" (*PII* 469). There are a number of similar passages in Lawrence's essay, for the essence of the piece is Lawrence's attempt to justify his killing of the porcupine in terms of his philosophy of a hierarchy of vitalism.

In all of his analyses of Lawrence and evolution, Ebbatson emphasizes the early fiction, character typology, and sexuality. He ignores almost entirely the centrality of food in the above quotation to concentrate instead on sexuality and mentions Lawrence's middle to late fiction only occasionally in passing. While I have no quarrel with much of what Ebbatson has to say regarding Lawrence and evolution, I wish to concern myself with the areas Ebbatson ignores or avoids: the centrality of factors such as food that are involved with physical survival, the narrative structure, and the middle to late fiction. I believe, as well, that Ebbatson errs in accepting everything Lawrence writes as equally authoritative and serious. Lawrence is sometimes deliberately polemical, and he resorted at times to a rhetoric of extremity in his non-fiction as well as his fiction. What he was serious about was employing the tools of Darwinism in an effort to reconfigure the course of his own work and the structure of his culture.

How, then, does Lawrence use Darwinian concepts to imagine the reconfiguration of his culture? The short answer is, with great difficulty. For an individual to thrive in the Lawrentian sense, society must be so constructed that it does not impede fulfillment. But for society to attain such a level of benign nurturance vis-à-vis its individuals requires that those who hold power recognize the problem and are motivated to bring about the necessary change, and people capable of such discernment will not be produced by society as presently constituted. This catch-22 leads Lawrence to envision a reordering of the social hierarchy in the first place but involves him in the further conundrum that if force is used to bring about the kind of society that will nurture individual fulfillment, the very use of force itself prevents that kind of society from coming into being. Lawrence tries in several essays of this period to distinguish between power and force because he realizes the fine line he is treading.[15] In place of the aristocracy of land and birth, Lawrence would establish one of vitality; instead of the hierarchy of a Kiplingesque racialism he would, despite occasional disturbing and offensive racist remarks, remind the white race of its weakness vis-à-vis the "primitive" races, even in one poem, "Tropic," imagining himself as a member of another race: "Behold my hair twisting and going black. / Behold my eyes turn tawny yellow / Negroid" (*Poems* 301).[16] Instead of what he views as "Cocksure Women and Hensure Men" (in the words of the title of an essay published in *Forum* in January 1929), he would reestablish what he thinks of as a healthier and more natural relationship between the sexes, the cocks and hens returned to their "rightful" places. The Nietzsche in Lawrence drives him to imagine an abolition of the old hierarchies for a new vitalistic one, so that ultimately the hierarchies of class and race become subsumed to the hierarchy of existence and survival.[17] In "Education of the People," Lawrence offers this organic image of hierarchy: "The populace finds its living soul-expression cumulatively through the rising up of the classes above it, towards pure utterance or expression or being." Such a rising up is compared to the natural flow of sap through the trunk of a tree, issuing in the end in "the perfect blossom," the "vital" individual, Lawrence's ideal leader (*P* 610).

Beer suggests that Darwin "eschews the simplicity of hierarchy" (100), that he moves away from the image of the medieval Great Chain of Being to the concept of a web. However, it might be argued that evolution, at least as it was popularly conceived, suggests a hierarchy just as strongly as does the Great Chain. The difference between the two is that the barriers in Darwin's hierarchy are permeable – and that was the crux of its revolutionary effect in class-bound England – while the Great Chain is conceived in terms of fixed links except perhaps at some future Judgment Day. No toad will ever become an angel, although the Elect may join the heavenly host on God's right hand after the Apocalypse. In a sense, Lawrence wishes to maintain the hierarchy of the Great Chain, or rather, Congregationalist that he is, he is incapable of conceiving of any other model of hierarchy, but his principle of ordering within the hierarchy is vitalism, not divine creation, and his Judgment Day is now. He is a sick man in a hurry.

We might say that earlier in his career the hierarchy of the Great Chain is implicit in Lawrence's work as the biblical metaphors of his childhood hold sway, metaphors that late in his life he looked back upon with nostalgia (in "Hymns in a Man's Life," for instance). We see the Great Chain at work in central passages such as the one where *The Rainbow*'s Tom Brangwen, at the wedding of his step-daughter, Anna, describes a married couple as an angel, thereby implying the force of human love and sexuality to create something transcendent or at least ascendant (*R* 129).[18] Where his granddaughter Ursula, at the end of the book, imagines a new dispensation in terms of God's biblical promise of a new start only after a deluge of destruction, the antediluvian Tom can imagine the fulfillment of the dream of aspiring to angelic condition, one rank above humanity, in marriage. In the course of but two generations, then, there is a felt need for a much more drastic or at least destructive solution to the problem of how to establish a social context that will nurture the coming into being of the fulfilled individual. But both views are contained within biblical terms of reference.

A further complication in considering Lawrence's imaginative reworking of hierarchy is that, even as he champions in the works of this period a new definition of aristocracy based on the vitality and leadership qualities of individual (male) human beings, Lawrence begins in earnest, as Norris argues, to question the basic right of the human species *qua* species to lord it over Creation. As he becomes more and more disillusioned with human conduct, and as he searches desperately for the right social-cultural formula, part of Lawrence's postwar strategy is to displace the priority of the human world with the "biocentric," that is, to undermine the long history of anthropocentrism in Western culture and its sanction in the biblical injunction by God to Adam to wield dominion over the natural world. The positive side of Lawrence's misanthropy is a lack of what we might term species narcissism that allows him to see, Wordsworth-like, into the life of things. Part of Lawrence's originality as a writer (and what he persuasively suggests is Cézanne's greatest strength as a painter)[19] was, from the very first, his ability to describe the natural and animal worlds in ways that convey a convincing impression of their quiddity. However, where this talent early in his writing career tends to be a symbolic adjunct to an unfolding human drama, more and more, as Lawrence's misanthropy increases, his x-ray vision is put at the disposal of a displacement of the human that tends to degrade human life itself. One example is Birkin's retreat – "naked among the primroses" (*WL* 106) – into the consolation of the vegetable world after being struck on the head by Hermione Roddice with a lapus lazuli paperweight in *Women in Love*. The implication is not only that the natural world may be consoling or that it contains features worthy of human emulation but that, in Birkin's mind at this point at least, the natural world is far preferable to the human: "why should he pretend to have anything to do with human beings at all? Here was his world, he wanted nobody and nothing but the lovely, subtle, responsive vegetation, and himself, his own living self" (*WL* 107). There is something disturbingly defensive about Lawrence's botanical orientation, his characteristic use – overuse in the case of *The White Peacock* – of the tree-branch-bud-flower-seed metaphor cluster as a touchstone for human life. Lawrence's identification with the process of entelechy and growth in the (immortal) vegetable world becomes, especially in the mid to late fiction, part of his mechanism for coping with his death anxiety, his impaired sexuality, his shrinking body mass, in short, his own decay. In the "Porcupine" essay, he declares that "Man, as yet, is less than half grown. Even his flower-stem has not appeared yet," and that "[e]ither he will have to start budding, or he will be forsaken of the Holy Ghost: abandoned as a failure in creation, as the ichthyosaurus was abandoned" (*PII* 471). In a companion piece, "Aristocracy," written at about the same time, when he was still attempting to justify his killing of the porcupine, Lawrence seems to identify his own life with the tree of Life and to be giving expression in projective form to his own fears of being consumed by consumption: "The world is stuck solid inside an achieved form, and bristling with a myriad spines, to protect its hulking body as it feeds: gnawing the bark of the young tree of Life, and killing it from the top downwards" (*PII* 484). A later example of Lawrence's biocentric displacement of human priority (this time in favour of the animal world) and the one used by Norris as a *locus classicus* is the story *St Mawr,* in which Lou Witt and her mother come to resent the exploitation of animals indulged in by the horsey set that makes up their class and seek to remove themselves entirely from society. St Mawr, the stallion, however, comes to be seen not only as worthy of respect as an animal but, in some ways, of more intrinsic value than Lou's husband, Rico, or any other human for that matter. In such radical ecologist moods, Lawrence is not content to engage in what Norris calls "the dismantling of an anthropocentric ontology" (170); he is happy to contemplate the extinction of the human species as an improvement, on the whole, for the planet. Such a view is unthinkable – at least in the form it takes – without the theoretical framework provided by Darwin.

To a great extent, Lawrence carried out his program of realigning the hierarchical values in his fiction through the use of survival and death as narrative tools. As we will see, Lawrence's choice of narrative device is overdetermined in that it comes to serve not only a conscious narrative strategy but also his own deep psychological need. Once we have learned in each of Lawrence's works who dies and why they die, we have gone a long way toward elucidating the narrative strategy he is using and where it ultimately comes from. For, unlike the work of many of his fellow moderns who had highly developed senses of ambiguity, Lawrence's fiction often displays clear winners and losers.[20] In Lawrence's first three novels, *The White*

Peacock (1911), *The Trespasser* (1912), and *Sons and Lovers* (1913), death comes as a culmination, the conclusion to a process that in some ways is presented as tragic. In this, Lawrence seems to be working through his debt to Thomas Hardy, arguably the greatest tragic novelist in the English tradition since Richardson. That such is not Lawrence's natural territory as a writer, however, becomes abundantly clear in *The Trespasser*. *Sons and Lovers* is a marked improvement in part because Lawrence no longer depicts death as necessarily tragic. Yet even so early, one may detect vague indications of the concept of death as a tool that becomes fully developed by *The Plumed Serpent*. As John Worthen has shown, the writing of the death of Mrs Morel in *Sons and Lovers* was excruciatingly difficult for Lawrence (*The Early Years* 451–5). Yet it strikes the reader now as inevitable and perfectly in keeping with the structure of the novel and with the description of Paul's growth as an artist and a man. Mrs Morel's death liberates Paul, but it comes at the end of the novel so that we are left with the impression that Paul's self-realization would have come at some point anyway. Paul resorts to euthanasia, not murder. Nevertheless, the pattern is set whereby the death of the other (especially the motherly other) is strongly connected to self-liberation. The theme of release upon the death of a parent is common enough, especially in Edwardian fiction (Trodd 52), but with Lawrence the release takes on ominous aspects. Paul Morel feeds his mother poisoned milk, thereby simultaneously commenting on the toxic nature of maternal nourishment (when administered, like Mrs Morel's medicine, in too large doses) and avenging himself in a symbolically appropriate way.[21]

In his fiction transitional to the leadership works, Lawrence most often envisions male liberation or fulfillment in such a way as to suggest that it is contingent upon the death of a female rival or of a mother figure. Jill Banford in *The Fox* is killed by a tree, representative of the natural world with which she cannot cope, so that Birkin's vegetative world is still friend to man, but it has now become enemy of woman. *The Captain's Doll*'s Mrs Hepburn is killed by a fall, symbolically negating as well as literally expunging the exalted position in which she has placed herself in her marriage. In both cases, Lawrence leads up to the woman's death imagistically with the same two motifs present in the scene in *Sons and Lovers*, food and illness.

Up to and including *Women in Love*, these kinds of vitalistic deaths would normally have been internalized into the death of the self or of an aspect of the self as a prelude to an internal rebirth. *The Rainbow* is replete with scenes that attempt to depict such a process. Beyond *Women in Love*, liberation seems to require what in "The Woman Who

Rode Away," in its most extreme form, is presented as actual human sacrifice that has no necessary internal correlative. The woman's projected death in that story may liberate a tribe of Amerindians in a collective sense, but there is no convincing reason for the woman to ride into such a situation to begin with. The protagonists in the leadership works generally have a highly instrumental attitude towards death as it becomes a political tool to forward a program of vitalism, although Kate Leslie in *The Plumed Serpent* represents a vestigial female resistance to that program.

In the transitional works, we see Lawrence beginning to alter his use of death in his fiction in keeping with his goal of "creating an ideology which was a prescription for survival achievable not instantaneously but over time" (Templeton 186–7). In many cases, death seems to become the central actualizing event and to carry heavy symbolic overtones. For one thing, it allows Lawrence to reshuffle the deck of hierarchy, to present us with a different concept of strength and weakness in character that the plot helps him to sort out. Ostensibly powerful individuals are, in the new hierarchy, often unable to survive, while those who appear powerless, like the prisoner of war Count Dionys in *The Ladybird*, turn out to be endowed with a less tangible but crucial sort of power.[22] Death, then, becomes a sign of a lack of vitalistic powers. However, it is important to note that Lawrence seems to fudge his logic here. In what amounts to a tautology, death through chance carries off the person unfit to survive, but often the only convincing proof of unfitness is the fact of death itself. Gillian Beer has pointed out the same circular structure inherent in the Darwinian doctrine of "survival of the fittest" itself (117), but the circularity for the evolutionists is deliberate, since the definition and only significant sign of fitness is and can only be survival, and no ideological scheme determines which individuals survive. Lawrence may be said to "inherit" from Darwin and Spencer the circularity of his argument at this point, but the result in his fiction is unfortunate. It has the sanction of the very evolutionary theory Lawrence claims to reject.

In a study of the modern short story, John Bayley discusses Lawrence's short fiction in a way that elucidates the ruthlessness and autobiographical elements of the period we are looking at. Bayley does not make the connection to Darwinian doctrines like survival of the fittest in any explicit way, but his observations point in that direction and take implicit account of the metafictional dimension to the ruthlessness of protagonists when they are "creating action or situation in the same way that the author does in creating the tale" (125).[23] The comments are worth citing at some length since they express what many readers of the fiction of this period sense without being able to articulate:

> If the hero of "The Fox" were a real murderer, who had engineered the girl's death deliberately to make it look like an accident, the story would have no significance. The point is that Henry *is* a murderer, in his own mind and desires, as Lawrence himself certainly was, and no doubt many of us are: a man who is quite ruthless in getting what he wants at the expense of other people, who, like Banford, can be felt to have no "real" existence. The lack of a proper Lawrentian existence in the frail and querulous Banford means that actual existence can be determined as an act of the will by the hero on behalf of the author. (125)

Throughout the *Origin*, Darwin uses the "analogy of husbandry" to press home his argument regarding nature's own means of selection and preservation (Beer 32). Lawrence applies the principles of husbandry, as it were, both fictionally and metafictionally, in his own work. Particularly in the transitional fiction, the issue of being a husband in both senses of the word is at least as important as the exploration of political issues. In *The Fox, The Ladybird, The Captain's Doll, The Lost Girl,* and *Aaron's Rod,* and in a number of short stories from the period, the two meanings of "husband," the Darwinian and the matrimonial, are played off against each other. A good husband in a marriage relationship, these texts argue, is able and willing to husband the human flock by removing its weak or unfit members and by returning the female of the species to her "rightful" place. The revisionary quality of Lawrence's work in this period parallels the content in terms of husbandry. The writer allows some ideas and characters to survive the conflicts he engineers, while others perish. It makes most sense, in the end, to speak of Lawrence's relation to Darwinian concepts within the context of his own evolution as a writer and of his own notion of narrative.

Beyond an examination of Lawrence's use of survival and death as fictional tools, it is instructive to consider the ways in which growth and transformation, two important concepts in the nineteenth century (Beer 104), operate in the work of both Darwin and Lawrence. Briefly stated, in Darwin's scheme growth is an ontogenetic process of little interest to species development except insofar as it determines whether an individual will survive long enough to pass on inheritable characteristics. Transformation, however, is the phylogenetic result of natural selection over time and so the focus of Darwin's study of natural selection in species. Growth and transformation are always important in Lawrence's exploration of individual development, but where growth by definition suggests expansion and largely controls the marriage novels, transformation may imply destruction of the old, a burning off of impurities, and is central to the leadership works. In the marriage novels, any rejection of growth in favour of transformation is confined, more or less, to Lawrence's revolutionary portrayal of individual character. What Lawrence wishes to do with Darwin's concept of evolution at this point is internalize it within the individual, thereby transforming the idea of species surviving through adaptation to that of the individual going through successive "rebirths" in order to reach full potential or, as Lawrence would call it, full flowering. The two "Widening Circle" chapters of *The Rainbow* suggest a still-positive view of expansion and growth. In *The Rainbow*, one could argue, Lawrence tries to make the individual both the generational means and the individual end of evolution. In any case, for Lawrence – and in this he for once remained consistent throughout his career – the individual is always the end – and sometimes the target. The individual can be the target because in the leadership works Lawrence's view of evolution, it seems, shifts in emphasis from the exfoliation of the individual to the necessity to destroy unhealthy tendencies, cultures, and individuals. The shift is incipient in the fiction transitional to the leadership period.

In the novella *The Captain's Doll*, growth is symbolized especially by the three thousand acres of an African farm that the protagonist hopes to travel to, an expansive image introduced only at the end of the story and never developed. Transformation is imaged by the moon (a continually changing orb), which conveniently "kills" Mrs Hepburn by causing vertigo, apparently, as she stands by an open window, an event that allows the eponymous captain to pursue the much younger and attractive Hannele.[24] Mrs Hepburn is expendable in evolutionary terms because she is seen by Lawrence as incapable of evolutive transformation. The earlier "fixity" that, at least according to Hannele's friend Mitchka, characterized the captain seems to die with his wife. But why other than for his gender he should be deemed more adaptable than she and thus be earmarked for survival is difficult to see. The attack upon the mother/wife figure that characterizes the fiction and essays of the period is the natural extension of the view that growth, even if desirable, is no longer possible because, for Lawrence, modern mothers, instead of providing the sustenance and scope needed for growth, offer corruption and confinement. If growth is no longer possible in a decadent society, then transformation becomes even more important than it would otherwise be, and it becomes urgent to deal ruthlessly with anyone who blocks it.

By the leadership works, however, even the death of selected individuals is no longer enough, nor is the transformation of those fit to survive. The now-moribund society must undergo a complete metamorphosis. Hence the need for a leader who will effect the required change through the doctrines of survival even as Lawrence metafictionally subjects his own fiction to transformation through ruthless selection. It is largely such uses of Darwinian concepts by Lawrence that are responsible for the unattractive quality of much of his fiction after *Women in Love*. But the tendency was certainly there all the time, as we can see in the early short story "A Modern Lover" (written in 1909 but first published posthumously in 1933), where we get a clear illustration of both the Lawrentian internalized death and the ugly side of Lawrentian transformation in the absence of growth.

Unlike the Brangwens of *The Rainbow*, the "modern lover" of the story, Cyril Mersham, is unappealing mainly because we get a sense of a transformation that involves not only a conscious destruction of his own earlier self but also a defensive, even aggressive, attitude toward anything that might remind him of that self. The general storyline is similar to the Clara and Baxter Dawes subplot in *Sons and Lovers*. Cyril first does all he can to break up the relationship between Muriel (the Miriam/Jessie figure) and Tom Vikers. However, once he can see his rival as an agent in his own further transformation, he begins to feel great sympathy for Tom and even a physical attraction. The idea of the self as constantly transforming – a caterpillar-pupa-butterfly development – is central to the story. Cyril realizes that "nothing is so hateful as the self one has left" (*CSS* 6). He imagines an upward path that has marked his intellectual progress. He has left in the wake of his own evolutionary development such authors as "Carlyle and Ruskin, Schopenhauer and Darwin and Huxley" (*CSS* 6). Even in relegating Darwin to his past, however, the hero unconsciously endorses a concept of individual evolution that is clearly Darwinian. Furthermore, this despicable young man seems to consider the members of the farming family with whom he was once on such intimate terms, as the young Lawrence himself was with the Chambers family, as somehow closer to lower forms of life than he is himself. If there is an ironic distance between the narrator and the perceiving consciousness of Cyril Mersham, it is difficult to detect: "Mersham watched them, as he had watched the peewits and the sunset" (*CSS* 5). He views his own soul as a seagull flying low over the waters, "stooping now and again, and taking a fragment of life – a look, a contour, a movement – to feed upon" (*CSS* 2). Lawrence thus (deliberately or inadvertently) reveals Cyril Mersham to be a predatory young man whose awakening and evolutionary development have come at the cost of his deepest relationships, just as in *Sons and Lovers* Paul's liberation will be at the cost of his mother's life and Miriam's happiness.

Such Darwinian tendencies in Lawrence's early writing develop by the time of *The Plumed Serpent* into something less subtle and much more alarming. Instead of earlier selves that must be left behind on the road to self-fulfillment, the corpses of enemies will be strewn about in the struggle for the Utopian society. The celebrated opening bullfight scene, the only portion of the book really worth reading, sets the stage for an entire society to be portrayed as a ritualized arena of the survival of the fittest. It is clear that part of what pushes Lawrence onto such a byway is the First World War, a particularly difficult time for him in several ways, as the writer himself and many of the commentators on his life and work have documented in abundance.[25] We might describe Lawrence's difficulties attendant on the war in terms of survival and fitness. Repeatedly and humiliatingly declared unfit for a military service that he saw as detestable in any case, cut off as a writer from his audience by the war-related suppression of *The Rainbow* and his subsequent difficulty in finding a publisher for *Women in Love*, periodically ill to an alarming degree,[26] Lawrence in the war years was marginalized. His place of residence for two years of the conflict at the western extremity of England is aptly emblematic of his position at the edge of survival both as an artist and as a living being. But from the beginnings of Lawrence's visceral reaction against the war and everything he felt it represented of a doomed and destructive culture to the "solution" of *The Plumed Serpent* there is a long, convoluted journey taken in the fiction. At the crossroads of that journey stand the transitional works I will be discussing.

As many commentators in the last two decades have noted, the major change in Lawrence's fiction as he moved from marriage to leadership is the portrayal of women, which operates under the shadow of what Lawrence, referring to Frieda in a letter to Katherine Mansfield, called the "devouring mother" (*L* iii 565). Although Lawrence's portrayal of women changed significantly, the devouring mother figure is a constant presence in his work. In fact, Judith Ruderman has argued that the concept of leadership itself and its causal factors were there in Lawrence's fiction from the very beginning and remained to the very end (7). While this may be true – and men do die "because of" women as early as *The White Peacock* and *The Trespasser* – it is also a fact that in the 1919–22 fiction, Lawrence attempts to deal with the devouring mother in a much more ruthless fashion than he had done to that point. Driving much of the aggression towards her, as we shall see, is an often covert or unconscious layer in the texts of the period that raises questions about the viability of the very narrative enterprise in which Lawrence is involved.

In her 1984 study, *D.H. Lawrence and the Devouring Mother: The Search for a Patriarchal Ideal of Leadership*, Ruderman argues persuasively that Lawrence's "ambivalence toward the 'devouring mother' was the prime motivating factor behind the works of his leadership period" (174). The oedipal patterns apparent in *Sons and Lovers* and to a lesser extent in Lawrence's other fiction are merely a veneer for the much more important pre-oedipal issues that engaged him throughout his writing career, and it is those unresolved problems that, in Ruderman's view, propelled Lawrence into his leadership period. At the heart of the problem is the universal parent-child dynamic whereby the child is torn between dependency and the need for independence: "The child's ... need for the mother's nurturing and protection stimulates a desire to return to the womb, but this regressive dependency means that the child will in effect die, having been annihilated by the devouring, engulfing mother" (9).

Lawrence's own extremely intense relationship with his mother both strengthened his dependence upon her (especially in times of severe illness) and exaggerated the effort required to gain a measure of independence from her. A juxtaposition of the letters in which Lawrence expresses, on the one hand, the realization that he could do nothing without a woman at the back of him (*L* i 503) and his candid admission, on the other hand, that he had great difficulty weaning himself from his mother when he was twenty-two years old (*L* i 527) tends to support Ruderman's contention that the pre-oedipal relationship is "more important to an understanding of [Lawrence's] works than the heterosexual, genital love relationship for which they are commonly known"(11). The mother-child relationship, far from fading in Lawrence's work after his mother's death and his elopement with Frieda, actually becomes more pronounced (12). I do not wish to rehearse at length Ruderman's attempts to link Lawrence's mother fixation to various important female figures in his life with whom he seems to have had an ambivalent attitude (people such as Ottoline Morrell, Margaret Radford, Mabel Dodge Sterne, and, of course, Frieda); suffice it to say that the common factor is an intolerable feeling of dependency at a time of vulnerability (illness, poverty, wartime constraint), a feeling that would inevitably recall the childhood love for and dependence upon the mother. What I would like to take from Ruderman is, first, simply that the mother-child relation is a central obsession behind much of Lawrence's fiction, and, second, that the feeling of dependency so basic to that relationship manifests itself in terms of nourishment and caregiving. I would add the qualification, however, that once those subliminal pre-oedipal patterns in Lawrence's work are discerned, what they lend themselves to most usefully is a fuller understanding of the relationships between Lawrence's adult men and women and not those between adults and children. In other words, it is important to understand Lawrence's pre-oedipal issues because and insofar as they complicate and problematize his portrayal of adult heterosexual relations. Ruderman's perspective sheds light on the tangled and obscure origins of Lawrence's grudge against women as it manifests itself in his fiction. His tool is Darwinian doctrine; his unconscious motive seems to be a childhood complex to do with dependency and survival.

Worthen has commented that the killing of Lawrence's cousin Walter Lawrence by Walter's father in 1900 was due to a disagreement over "position and priority" (*The Early Years* 48). But, by Worthen's own account, the row was most immediately about food. When Lawrence came to re-enact the quarrel in his play *A Collier's Friday Night* (written in 1909), he emphasized the importance of food. In the play, the father complains to his wife: "You can get things for them [the children] – you can, but you begrudge me every bit I put in my mouth" (*Plays* 67). Food and health rather than sex are the keys to individual rather than species survival, and in the fiction of the transitional period I want to look at we shall see that beneath the overt concern with sexuality (Ruderman's "oedipal overlay" [8]) there is an agonized obsession with food and other survival necessities (the "pre-oedipal conflicts") that makes a Darwinian perspective in terms of fitness and survival not only appropriate but rewarding. Even the attitude to violence in the writing of the period may be linked to "the child's fear of being destroyed by the mother," which "leads to a rage to destroy her – to commit matricide" (Ruderman 21). Like other terms in the battle for survival, the violence seems to take on a biocentric justification as human struggles are resolved with reference to the animal world "in which animals kill – but do not murder" (Ruderman 184).

Given the origin of the leadership issue in Lawrence's attempts to counteract an infantile feeling of dependency, there is a fundamental irony in the leadership period as a whole in that Lawrence ends up advocating the same kind of dependency between leader and follower that he condemns between mother and child, indeed suggesting that the former can be an antidote for the latter. What Ruderman calls "the jarring discrepancy between means and ends" in the leadership works is the proposition that the concept of leadership requires the "same abnegation of individual personality as the mother-worship it replaces" (35, 171).

It is important to be clear about just what Lawrence's concept of leadership was and what it was not. Several letters are often cited in definitions of Lawrentian leadership, and with good reason. In a July 1918 letter to Cecil Gray, just prior to the period we will be looking at, Lawrence seems to see the human race in static terms and suggests how it is to be governed: "The chief feeling is, that men were always alike, and always will be, and one must view the species with contempt first and foremost, and find a few individuals, if possible – which seems, at this juncture, not to be possible – and ultimately, if the impossible were possible, to *rule* the species" (*L* iii 262). While most commentators tend to ignore or downplay the qualification "which seems ... not to be possible" and one that follows it in the letter, "they can only be ruled as they are willing to be ruled," it is also true that fiction – especially the kind of compensatory fiction Lawrence engaged in without apology – often enacts what seems not to be possible in real life. Towards the end of the period under consideration, in a letter to Mabel Dodge Sterne in April 1922, Lawrence warns the American: "I don't believe either in liberty or democracy. I believe in actual, sacred, inspired authority: divine right of natural kings; I believe in the divine right of natural aristocracy, the right, the sacred duty to wield undisputed authority" (*L* iv 226). The rub was that even in his speculative fiction Lawrence never could convincingly portray those few individuals fit to rule, or, if such individuals could be portrayed, they seemed perversely to resist the inspiration to don the sacred mantle of authority. For the most part, they are the paradoxical creatures Vivas calls "misanthropic saviour[s]" (79) and simply want to be left alone.

Barbara Mensch argues that "a possible hypothesis" for the leadership novels is that "they chronicle a change in Lawrence's art from a social to a continuously more individual psychological perspective" (24–5). Although Mensch may have a point when it comes to a work like *Aaron's Rod*, I believe that, generally speaking, just the opposite is the case. In "Education of the People" (written in late 1918), Lawrence declares: "though man is first and foremost an individual being, yet the very accomplishing of his individuality rests upon his fulfilment in social life. If you isolate an individual you deprive him of his life: if you leave him no isolation you deprive him of himself" (*P* 613–14). Naturally, there are times, such as the *Dies Irae* dramatized in *Women in Love*, when society is so destructive of individuality that the only recourse is a retreat from it, as the resignations of Birkin and Ursula from their employment suggest.

At times in Lawrence's fiction, the much-vaunted championing of individualism seems forced. When the prose rises to incantatory rhythm and realism gives way to symbolism we do well to ask what ideological engine is propelling the narrative at that point. In his *D.H. Lawrence: History, Ideology and Fiction* (1982), Graham Holderness asks just that. His answer is that, after *Sons and Lovers* and until *Women in Love*, Lawrence abandons the realistic mode and the tragic novel for a myth of individualism and organicism that finds its most eloquent expression in the very novel in which incantatory rhythm and symbolic presentation play so crucial a part, *The Rainbow* (174, 180). In Holderness's materialist reading, Lawrence returns to a more realistic and tragic view of history in *Women in Love* but eventually to the mode of *The Rainbow* again in *Lady Chatterley's Lover*. Unaccountably, the entire leadership period of Lawrence's fiction, the phase in which he consciously attempts to explore the issues of social organization and ideology, constitutes an absence in Holderness's own text. He simply does not deal with it, perhaps because, as the texts are realistic yet not tragic, this would upset his thesis.

Although Lawrence's revolution is to be a biological one in some sense, he is also deeply committed to reforming society, as Raymond Williams once observed (300).[27] The trouble is he has no idea of how to go about it. As a consequence, he leans heavily upon such thinkers as Schopenhauer, Nietzsche, Ouspensky, and Trigant Burrow in his search for an ideal alternative to modernity. It is especially when he is parroting such thinkers, though, that his own metaphysic becomes confused.[28] The influence of Darwin, Spencer, Huxley, and Haeckel, on the other hand, remains largely self-reflexive in the sense that the writer applies them chiefly to his own narrative method. It is Lawrence's narrative strategy as much as his new philosophical interests that leads him into the cul-de-sac that *The Plumed Serpent* more than any other text represents.

In many of the transitional works, Lawrence tends to juxtapose Darwinian ideas with the imagery of awakening (which he made into what amounts to his fictional trademark), the appeasement of hunger, the curing of illness, and/or the breaking out of confinement. A condensed example of the collocation may be seen in the history textbook commissioned by Oxford University Press, *Movements in European History* (1921), where Lawrence is obliged by the nature of his subject to consider the larger collective view rather than the merely individual. Touching upon the "rebirth" that characterized the Renaissance, Lawrence comments that where in the Middle Ages "man was alive, but blind and voracious," in the Renaissance, he "awoke," and the human spirit became "like a butterfly which bursts from the chrysalis into the air," leaving behind "[t]he narrow, devouring little world of the caterpillar" (*MEH* 157). Sleeping and waking even in this collective sense suggest a transformation in humanity, a renaissance or rebirth as spectacular as the caterpillar's into a butterfly. The enormous hunger ("voracious," "devouring") that characterized the earlier stage of development has disappeared only because there has been a collective awakening or transformation and a breaking out of the confinement of the cocoon. The obvious analogy is with the child's individual quest for independence from the mother that nurtured him as an infant, when he was "blind and voracious," and who continues to confine him in a cocoon of love and protection. As Lawrence begins to mythologize his own relationship with Frieda after *Sons and Lovers*, the motif of awakening becomes central within the frequently used fairy-tale intertext of Sleeping Beauty. For example, in the "First Love" chapter of *The Rainbow,* Ursula is described in early contact with Skrebensky as waiting "like the Sleeping Beauty in the story" (*R* 278). The interaction of the Sleeping Beauty frame of reference with the deeper anxieties of the mother-child relationship serves to complicate the transitional fiction and in particular the textual attitude toward women.

The fact that Darwinian concepts appear in Lawrence metafictionally is further evidence that evolutionary concepts are crucial to an understanding of his narrative strategy. In a letter to Waldo Frank in the midst of the war (July 1917), Lawrence indicates that, although *The Rainbow* was written as a pre-war work, from his present perspective, he feels that the novel in revision became one that "does actually burst the world, burst the world-consciousness in every individual. – *What I did through individuals, the world has done through the war*" (*L* iii 142, my emphasis). Lawrence is clearly moving toward an understanding of his own work as an arena in which the events of the world are reflected symbolically through his characterization and fictional action. It is a short journey from there to a view of his characters as subjects in Darwinian experiments that test survival skills. In "The Novel," Lawrence states that the novel must be quick, "interrelated in all its parts, vitally, organically" (*PII* 422–3). As an organic entity, the world of the novel has some of the characteristics of a garden (if not a jungle): "Now in a novel there's always a tom-cat, a black tom-cat that pounces on the white dove of the Word, if the dove doesn't watch it" (*PII* 418). It is subject, in other words, to the doctrine of survival of the fittest. At times, Lawrence appears capable of thinking of his works not only as organic entities but as individual, living beings, as is evident, for example, in a letter dated 23 April 1915 and written to his agent of the time, J.B. Pinker. Lawrence is warning Pinker that the novel he has just written, *The Rainbow,* may cause worries about obscenity. He is prepared, he says, to consider removing small bits of it, but cannot excise whole pages or even paragraphs: "You see a novel, after all this period of coming into being, has a definite organic form, just as a man has when he is grown. And we don't ask a man to cut his nose off because the public won't like it: because he must have a nose, and his own nose too" (*L* ii 327). True to his principles, Lawrence's usual method of revising even long novels was to begin afresh and write an entirely new version. Although noses were not the anatomical part that usually upset the readers of Lawrence's fiction or viewers of his paintings, the association of the organic unity of the human (male) body and the novel is significant. In his introduction to a late volume of poetry entitled *Pansies* (1929), each poem is a flower and each flower a thought (pansies/*pensées*). A poem is a thought "which comes as much from the heart and the genitals as from the head. A thought, with its own blood of emotion and instinct running in it" Like all organic things, these thoughts, these poems, have both an individual existence and combine into a collective existence, "a complete state of mind." The poems, then, are not only like flowers, are not only replete with blood and instinct, they, in effect, become little creatures: "each thought trotting down the page like an independent creature, each with its own small head and tail, trotting its own little way, then curling up to sleep" (*Poems* 417), somewhat like T.S. Eliot's feline yellow fog in "Prufrock." If thoughts be such creatures, they may surely be culled by their keeper. We might even view the evolutionary perspective as the prime agent in the very act of fictional recapitulation and revision that went on in the works transitional to the leadership novels, in the three *Ladybird* novellas, for instance, which are all revisions of earlier short stories.

Worthen's remark that there is little trace of Darwin in Lawrence's writing in spite of the novelist's early enthusiasm for the evolutionists (*The Early Years* 179) is true only in a narrow sense, for Darwin's influence is discernable in Lawrence's portrayal of character and in his own revisionism. After *Women in Love*, he internalized the doctrines of a Darwinian reading of nature, and evolutionary concepts, especially the idea of survival, came to play a crucial role in the very structure of Lawrence's transitional fiction.

2 Food and Illness: Survival in the *Ladybird* Novellas

The Ladybird, *The Fox*, and *The Captain's Doll* were three companion novellas that Lawrence wrote late in 1921 and that he wished to keep together in one volume.[1] Precisely because each one is short enough for Lawrence to manipulate and yet long enough to be exploratory, they represent, more than any other fiction of the 1919–22 period, the transition Lawrence was trying to effect between his concern for marriage (and all it connotes in Lawrence) and for leadership (with all its attendant baggage).[2]

In many ways, the three novellas are typically Lawrentian. There is in each the "archetypal" male "gypsy"/gamekeeper figure who enters the world of a woman sleepwalking her way through life the way Frieda liked to suggest she had been living in her married life with Ernest Weekley: "I was living like a somnambulist in a conventional set life," she wrote in her memoir of her second husband, "and he awakened the consciousness of my own proper self" (3). There are the natural landscapes described with a characteristic vividness. There is the insistence, so common in Lawrence's mature work (more precisely after *The Rainbow*), that the woman must submit her will to the man in a heterosexual relationship. However, there is also much that is unusual for Lawrence in these three works, much that strongly suggests an uncertainty, a groping for rather than an advocacy of a new way, and there is even at times blatant inconsistency. That he returned in each case to a much earlier short story as the basis of his novella lends credence to the concept of a revisionary effort.[3] Leo Gurko's description of the three works as "the supreme staging ground for Lawrence's ideas at a turning point in his life" (182) may be a trifle exaggerated, but certainly these novellas are crucial to any understanding of Lawrence's "turn" from marriage to leadership.

Although Lawrence suggested that *The Fox* belonged to the old world while *The Ladybird* had the "quick of a new thing" (*L* iv 447), the three works do, in fact, have enough in common to justify his decision to juxtapose them within the covers of one book. For one thing, each novella attempts a redefinition of love that implicitly repudiates Lawrence's initial foray into the question of love in the Hardy essay's contrast between Love and Law.[4] In each, one kind of love is explored and rejected in favour of Lawrence's new conception of what love means, without that rejection having the full weight of authorial imprimatur. Although homoerotic love had offended Lawrence when he visited Cambridge as Bertrand Russell's guest, he explored it fictionally in a variety of ways before the period of the novellas: in Ursula's budding lesbian relationship with Winifred Inger in *The Rainbow,* in the suppressed prologue to *Women in Love,* as well as in the relationship between Birkin and Gerald in the latter work. In *The Fox,* lesbian love is suggested in the domestic arrangement between Banford and March. Christian love as Lawrence saw it is vehemently rejected in *The Ladybird,* and "modern" love and marriage as he conceived it in *The Captain's Doll* is shown to be a cul-de-sac. The fact that, in essay after essay late in his career, Lawrence denounces the wrong kind of love suggests that the three novellas merely initiate a never-to-end intellectual struggle rather than find a viable solution. "When it comes to women, modern men are idiots," he writes in "Give Her a Pattern" (1929). "They don't know what they want ... They want a cream cake that is at the same time ham and eggs and at the same time porridge. They are fools" *(PII* 538). In "Do Women Change?" (1929), he claims that "[m]odernity or modernism isn't something we've just invented. It's something that comes at the end of civilizations." In such decadent cultures, women are "modern," which means "[t]hey were smart, they were *chic*, they said cut-it-out, and did as they jolly well pleased" (*PII* 540). In "Men Must Work and Women as Well" (1929), Lawrence goes so far as to suggest that "[t]he sexes can't stand one another ... when a girl must be a flat, thin, bodiless stick, and a boy a correct mannequin, each of them abstracted towards real caricature" (*PII* 589–90). One recalls the totemic image of the doll in *The Captain's Doll* in the image of the mannequin. The redefinition of love in the volume of novellas is striking given that, relatively speaking, in much of his fiction and non-fiction in the decade following *Women in Love,* Lawrence attempts to move away from the terrain of man-woman relations and to focus more on the new interest in leadership and comradeship. But it is precisely this return to his older concerns in the *Ladybird* novellas that shows Lawrence's determination to make his earlier vision somehow square with his new (even if the peg he is using is a round one).

Closely associated with the redefinition of love in these stories is Lawrence's portrayal of the English class system and his implicit argument that a new hierarchy of vitalism must take its place for the sake of his culture's very survival. In *The Fox*, for instance, the two women who attempt to survive during wartime by raising hens fail miserably not only because of their gender, something we might expect from the Lawrence of this period, but also because the person who controls the farm, Jill Banford, lacks the requisite vitality (as suggested, among other things, by the fact that she requires spectacles to see properly). That the keen-eyed Henry Grenfel is far better able to deal with the crises that occur in the natural course of things on a farm than are the women is, correspondingly, due not only to his gender but also to his high ranking on the scale of Lawrentian vitality. Jill's jeering remarks about his lower-class behaviour underscore Lawrence's point (despite his own occasional tendency to toady to the upper class) that while traditional class consciousness may determine who has the better manners at the tea table it cannot possibly determine who will survive and who will not. Similar hierarchical redefinitions take place in the other two novellas, as we shall see.

Each of the three novellas portrays events during or immediately after the war and presents a soldier as its protagonist, perhaps because of Lawrence's conviction, as expressed in "The Risen Lord," for example, that the war weakened the hold of women over men: "During the war, the man who suffered most bitterly suffered beyond the help of wife or mother, and no wife nor mother nor sister nor any beloved could save him from the guns" (*PII* 572). Henry Grenfel is a recently demobilized soldier who had joined the armed forces at a young age and run off from Bailey Farm and the authority of his grandfather in order to do so. Both Basil Apsley, the aristocratic English officer married to Lady Daphne in *The Ladybird*, and his enemy counterpart, Count Dionys Psanek, are wounded and captured in the war, which ends during the course of the story. Alexander Hepburn in *The Captain's Doll* is part of the allied force occupying parts of Germany in the first few years after the war according to the terms of the Treaty of Versailles.

Given his visceral hatred of the First World War, Lawrence's use of soldiers as protagonists may seem strange, but he appears to have had an abiding fascination with military life, perhaps because of the male comradeship it seemed to offer. In *Mr Noon*, there is an entire section in the untitled chapter XIX where the only tension in the narrative is produced by the possibility that Gilbert Noon, described with some ironic distance by the narrator, may run off with the soldiers he so obviously admires: "He felt sick of women who talked and discussed and had privileges, of a theoretic life. Oh if only he was with the common soldiers! – the man's reckless, manly life of indifference and blood-satisfaction and mental stupidity" (*MN* 280–1). Nevertheless, in the three novellas and elsewhere in Lawrence's writing – most notably in the portrayal of Anton Skrebensky in *The Rainbow* – soldiering has a far from honourable place. The narrator of *The Fox* insists that Henry is a hunter, not a soldier; Hepburn resigns his commission when he gains financial independence upon his wife's death; and both Dionys and Basil are soldiers no more at the end of hostilities. Of course, war makes soldiers as much as soldiers make war, and none of these men, unlike Skrebensky, defines himself as a soldier. On the other hand, it suits Lawrence's purposes to have soldiers – professional or not – for his male protagonists in these tales in which the question of leadership plays so large a role. In his essay "The Two Principles," Lawrence speaks of the lumbar and sacral ganglion centres of the body, the seats of "the great sensual will to dominion. From these centres the soul goes forth haughty and indomitable, seeking for mastery. These are the great centres of activity in soldiers, fighters: as also in the tiger and the cat the power-centre is at the base of the spine, in the sacral ganglion" (*PII* 237). All three male protagonists in the novellas are, in fact, likened to cats (*Fox* 38, 92, 181). In fact, a fourth important similarity among the three novellas is their use of animals. We may see this development as part of Lawrence's biocentric challenge to an anthropocentrism threatening the natural and human worlds alike. He portrays the animals in these stories as being at risk of extinction (the fox), or under threat by modern technology (the mare and her foal swerving from the advance of the motor car in *The Captain's Doll* [*Fox* 127–8]), or associated with an extinct civilization (the scarab beetle of *The Ladybird*).[5] The dangers of war and the appearance of endangered animals are two textual manifestations of what remains largely a subtextual focus on the savagery of fitness and survival.

There is a certain amount of identification or totemic association between a number of the characters in the novellas and the animals who are at risk. Even Lady Beveridge in *The Ladybird*, who, from one point of view, is meant to represent a passé ideal of Christian love (much like Mr Crich in *Women in Love*), is the recipient of an obvious measure of the narrator's sympathy when she is associated with a doomed social system and, through the Virgin Mary imagery, with the Mary beetle that comes to symbolize the power of the "natural" aristocracy: "But during the last years of the war power slipped out of the hands of her and her sort … Then it seemed as if the many swords had gone home into the heart of this little, unyielding Mater Dolorosa. The new generation jeered at her" (*Fox* 157). Although Lawrence wishes to contrast Lady Beveridge's Christian charity and love with the Count's doctrine of power, his chest wound obviously links him with the English aristocrat and her pierced heart. They are both aristocrats, but only the Count is one in the double sense, by virtue of his birth and of his vitalism. Lawrence once Christologically termed the war itself "the spear through the side of all sorrows and hopes" (*L* ii 268).

Finally, in all three novellas, there is a strong feeling of confinement or constriction that is warded off to a greater or lesser degree by the protagonists. In *The Ladybird*, for instance, Dionys's confinement is a function both of his war wounds and his capture as an enemy officer in war. It is important to note that his resistance to the halter of confinement leads directly to his rhetoric of destruction, perhaps in a simple emotional compensation for his lack of real power in a foreign country or even in Austria, where his "castle has been burnt down in a riot" (*Fox* 168). In the case of Lady Daphne, the confinement takes the form of a repression of her feelings and desires so that Dionys's liberation at the end of the war coincides with Daphne's inner release. In *The Fox*, Henry and the fox he so strongly identifies with feel hemmed in by civilized society. At the end of the novella, he looks beyond the tight network of English hedges to a life in Canada, a country he presumably associates with wide open spaces. The feeling of confinement surfaces in *The Captain's Doll* in a number of ways. The heroine, Hannele, like her fellow Germans, is under the authority of an occupying military force, although the immediate threat in her case is not imprisonment but banishment from town. After the death of Mrs Hepburn, her husband, the eponymous captain, describes his wife to Hannele in terms of a caged bird. Hepburn himself seems to feel stifled in some indefinable way. His astronomical hobby of looking out at the moon and stars suggests a desire to escape from confinement, while the expansive image of a large farm in Africa appears towards the end of the novella as his destination.

We may also relate the concept of confinement to that of growth, in some ways its conceptual opposite. In the first chapter I suggested that Lawrence, in shifting his focus from marriage to leadership, was moving his emphasis from growth to transformation. In these three novellas, we see that change at work. When growth is no longer possible due to confinement of one kind or another (and the usual kind is associated with a mother figure) then the emphasis falls on transformation. But for transformation in the individual to be possible, society must be reformed or the possibility of social renewal must at least be imaginable, as at the end of *The Rainbow*. Enter the Lawrentian leader figure as the only possible catalyst for a social reconstruction that will take into account the needs of the individual.

There is a biographical element to the motif of confinement, as there invariably is in Lawrence. Lawrence is looking back to the war, when he was not only under suspicion as a spy but under orders regarding his movements and place of residence for the latter part of the war. Like Henry Grenfel, he must have felt during those years that "England was little and tight" (*Fox* 38).[6] Such "crowding" must inevitably lead to the imagining of an increase in evolutionary – even Malthusian – pressure since the urgency with respect to survival increases with competition.

Beyond the fictional portrayal of the battle for survival in the novellas, there is a metafictional struggle going on as Lawrence, typically, attempts to sort out his own ideas by subjecting them to the pressure of self-criticism dramatized fictionally. In "The Thimble," the story that Lawrence used as the initial basis for *The Ladybird*, a husband badly mutilated in the war comes home to his wife, who is just recovering from pneumonia. As she awaits his arrival, the wife discovers an old, ornate thimble in the sofa she is sitting on. Somehow the object comes to represent to the husband the obstacles standing in the way of their rebirth as individuals and as a couple, and he tosses the thing out the window onto the street where it disappears. That is how the tale ends. In *The Ladybird*, the thimble becomes the central symbol in a story that uses many of the elements of the earlier tale but transforms them radically. It is as if the same thimble that had been discarded in the earlier story is retrieved for use in the novella because Lawrence himself has discovered that its work is not done or that he had misunderstood the work that it had to do. Such intertextual references are not uncommon in the transitional period, as Lawrence gropes his way forward toward the leadership works. In a letter to Earl Brewster, Lawrence refers to his story "The Fox" as if it were itself an animal. He has, he writes, "put a long tail" to what was "a bobbed short story. Now he careers with a strange and fiery brush" (*L* iv 126). Lawrence identifies his own endangered career with the careering of the fox that is hedged in by English fences and faced with extinction, much as he had felt himself to be in Cornwall during the war. Henry Grenfel is, in fact, Cornish. What is at risk, in the end, is Lawrence's own conception of reality and, thus, the kind of fiction he is capable of producing. He feels it necessary in 1921 to add a "fiery brush" to "The Fox" so that, unlike the animal *of* the story, the fox that *is* the story itself can survive his new norms. If Henry "saves" March in *The Fox*, what "saves" Lawrence's story for his new ideology is the "tail" he refers to metafictionally as an appendix to the story. Fictionally, the tail of the fox in March's dream wounds her in a way that adumbrates her sexual conquest by Henry and the destruction of something inside her whose external analogue is her friend Jill. Metafictionally, Lawrence is suggesting not only that his art must incorporate new ideas (the fiery brush) in order to survive but also that his own survival as an artist depends upon subduing his earlier view of life and, most particularly, the female component in himself. From an aesthetic point of view at least, he is dead wrong, as he eventually comes to realize himself.

THE FOX: FELLING THE PROBLEM

The Fox, one of Lawrence's best-known stories, tells of the efforts of two women in their thirties, Jill Banford and Nellie March, to run a farm on their own during the war. They are having numerous difficulties when, at war's end, a youth named Henry Grenfel, who used to live on this same farm with his grandfather, returns and begins to help them with the farm work in return for a place to live during a leave from the military. A fox has proved to be one of the more intractable problems faced by the women in their efforts at husbandry, and March has encountered the creature on several occasions in ways that suggest a mesmeric attraction based on her own unacknowledged instincts. When Henry arrives, he replaces the fox in March's semi-conscious awareness, and he, on his part, proceeds with an impulsive plan to marry March and wrest the farm from Banford's control. (Her father had financed the enterprise.) The conflict culminates in Henry felling a tree in such a way that it strikes and kills Banford, and the novella ends with Henry and March looking toward an uncertain future in Canada.

I begin with *The Fox* since the pattern I wish to discuss is more clearly present in this novella than in the other two. That pattern, as I elaborated it in the first chapter, consists of Lawrence's attempt by means of evolutionary ideas, especially the concept of survival, to redefine the hierarchical organization of English society in terms of vitalism, an attempt that leads to an inevitable and fatal digression into the area of parent-child relations. By emphasizing certain scenes and recurring elements in *The Fox*, one may draw attention to the concepts of survival and fitness, concepts that have been neglected critically.

Each of the three novellas has a key scene or description that, upon reflection, seems incongruous. Investigating the incongruity can be a lever to pry open the vault of Lawrence's failed narrative strategy in this period of writing. In *The Fox*, the discordant scene is a description that does not appear in the earlier version of the story (in which, in fact, the fox is not killed) of Nellie March examining the fox that has been killed by the Lawrentian hero, Henry Grenfel, and hung up by its heels in the shed. She is fascinated by the creature's "white and soft" belly and its "full and frictional" brush, obviously phallic in connotation, that she slowly strokes (*Fox* 41). So far, Lawrence's description in this scene is perfectly in keeping with his obvious intentions regarding the fox's symbolic value in a story of awakened female sexuality. He adds a paragraph to the description, however, that does not seem to have any apparent function in the story. March moves from stroking the fox's belly and tail to examining its head as Henry approaches and Banford walks off:

> She was wondering, wondering, wondering over his long fine muzzle. For some reason it reminded her of a spoon or a spatula. She felt she could not understand it ... Wonderful silver whiskers he had, like ice-threads. And pricked ears with hair inside. – But that long, long slender spoon of a nose! – and the marvellous white teeth beneath! It was to thrust forward and bite with, deep, deep into the living prey, to bite and bite the blood. (*Fox* 41)

It is not that the passage, on first reading, is terribly obtrusive. In fact, it is fairly typical of the Lawrentian version of modernist realism fraught with symbolic overtones. But it is often when Lawrence is being typically "Lawrentian" that we ought to ask what he is doing in his narrative, for perhaps then he has for the moment ceased exploring and fallen back upon his own compulsions unawares. There is a similar passage in *The Captain's Doll* that seems less out of place in the context but whose imagery is rather startling. A fox, after all, does bite and bite deeply; it is more remarkable when water is described as "rushing like a beast of prey," as it is at one point in *The Captain's Doll* (*Fox* 129) where the scene encompasses more literal "ice-threads" than the fox's whiskers here. There, the mountain valley encountered by Hannele and Hepburn on their way to the glacier has hair like the fox in the novella presently under discussion ("with black trees like hair flourishing in this secret, naked place of the earth"), and the "insatiable" water bites like some beast of prey:

it is "like the steel edge of the wedge, the terrible tip biting in into the rocks' intensity" (*Fox* 129).

On further consideration, what is odd about the description in *The Fox* is the recurrent associations with food and eating: the spatula and spoon, implements of cooking and eating, the teeth, the biting, the blood. Lawrence has not prepared us for the equation of awakened sexuality and eating. In relation to March, the fox has suggested sexuality not because it is a predator primarily but because it is natural, wild, and not to be repressed. The incongruity in this passage is part of a pattern in the novella that functions to undercut Lawrence's apparent narrative strategy. Wishing to suggest that health (including sexual health) is possible only with the aid of male leadership, Lawrence is unable to prevent the motif from sliding over into a competing pattern of pre-oedipal concerns that interfere with the former. In fact, the presence of food and eating in *The Fox* is as pervasive as it is subtle. It seems to suggest a compulsion at work that appears at least in part to be beyond authorial control.

We learn within the first few pages of the novella that, at this point in England, "[f]ood was scarce and bad" and that feeding the chickens is proving to be a great difficulty for the two women (*Fox* 8). At this juncture in the story, the spell of the fox and what it represents seem to compete in March's consciousness with food. At supper, she can hardly keep her thoughts on her food because she is preoccupied with the animal (*Fox* 11). After Henry arrives and the women are convinced that he is harmless, they offer him tea, but because "the larder was bare," they have only bread, margarine, and jam to serve (*Fox* 15). They do their best: "Banford flew into the village on her bicycle, to try and buy food. She was a hospitable soul. But alas, in the year 1918 there was not much food to buy" (*Fox* 20). In fact, it soon becomes clear that, although Henry has a large appetite, his arrival results in there being more rather than less food available on the farm. He is much better able to provide food for the three of them than the two women were for themselves alone. He not only kills the fox that has been depleting their stock of chickens, he also shoots a rabbit and duck, making "a great addition to the empty larder" (*Fox* 21).

Most of the confrontation scenes in the novella occur during mealtime: Henry's first manipulation of the women to allow him to stay on the farm for the duration of his leave; March's defence of Henry's coming to tea in his shirtsleeves, which causes Jill to feel "her food stick in her throat" (*Fox* 28); the announcement to Banford that Henry and March are to marry, a communication that prompts Banford to "put down her knife, out of her thin, delicate fingers, as if she would never take it up to eat any more" and to turn aside from her plate, "as if the sight of the food on the table made her sick" (*Fox* 34); and Henry's discovery of March at tea in a dress: "Banford, at the other end of the table, said not a word, but piggled with the sardine on her plate" (*Fox* 48). As Hepburn in *The Captain's Doll* curses the affected mountaineers he encounters in Austria with a wish to "make them all stop up there, on their mountain-tops, and chew ice to fill their stomachs" (*Fox* 137) so Henry wishes to make the flower-laden Banford eat flowers for her tea: "'And I'd give them you for breakfast again, I would. I'd give you flowers. Nothing but flowers'" (*Fox* 45). He feels that March would have "soft woman's breasts" while Banford would have "tiny iron breasts" (*Fox* 48). In Banford's apparent inability to eat, Henry's choice of mate leads to Banford's symbolic death even before her actual demise. The crucial point here, though, is that it is Lawrence's use of food as a motif to suggest quickness or survivability that triggers the parent-child paradigm his work of this time is obsessed with.[7]

Indeed, although March and Banford develop something very much akin to a spousal relationship within their community of two, with respect to the external world there is more of a parent-child alignment at issue, something that most critics of the story have failed to remark upon.[8] The animals, like children, must be fed and looked after. Like unruly offspring, they refuse to sleep at the wonted hour when Daylight Saving time is introduced, and "there was no peace till they were shut up and asleep" (*Fox* 8). Initially, Banford is happy to "mother" Henry as she might her own younger brother (*Fox* 16, 19). In fact, when Henry unexpectedly cycles back to Bailey Farm from his military camp just prior to the climax of the story, Banford's father at first ironically mistakes the figure in the distance for "one of our boys" (*Fox* 61). Henry is young, a good ten years younger than the two women and even more years their junior in the earlier version of the story.[9] Lawrence writes that "[h]is eyes were wide and childish, with thinking about things" (*Fox* 20). He is happy enough to be "mothered" his first evening on the farm, but very quickly he is providing food for himself and the other two. This begins a symbolic reversal in the parent-child alignment. Clearly, then, we are in the familiar Lawrentian territory of the mother-son-lover triad or, perhaps more accurately, the "devouring" mother-child dyad. In fact, along with Henry's obvious efforts to pry March loose from Banford's hold in an odd sort of sexual rivalry, there is a more subtle effort going on to invert the parent-child relationship: "What if she was older than he? It didn't matter. When he thought of her dark, startled, vulnerable eyes he smiled subtly to himself. He was older than she, really. He was master of her" (*Fox* 23). That Henry is so much younger than

March merely serves Lawrence to drive home his point that mastery is not a matter of age (or social standing); it is a matter of vitality. March's objection to Henry's marriage proposal – "'What nonsense! I'm old enough to be your mother'" – is swept aside by Henry as irrelevant: "'What is age? What is age to me? And what is age to you! Age is nothing'" (*Fox* 25). However, if age is nothing, a pert reader who confuses Henry's pronouncements with his creator's sentiments might ask the author why he would make Henry so much younger than the women, a narrative strategy that clearly suggests that age is very much something. Even the otherwise unaccountable presence of Jill's parents during the tree-felling scene might be explained as part of this pattern of parent-child inversion, for they are brought into the story for this one scene only, to witness their daughter's death, and the scene and the novella would work perfectly well without them. Why does Lawrence bring them in? There is a biographical parallel to the arrival of the Banfords at their daughter's home (Nehls i 465–7), but Lawrence seldom used the events of his life (in contrast to the more amorphous *emotions* of his life) as elements of his fiction without a clear idea or at least intuitive understanding of their function in his art. The presence of the two older people as witnesses to their daughter's death may function to underscore Henry's inversion of hierarchy, since the reversal of the normal order of death of parents and child may be taken as its analogue. More generally, as in *The Ladybird*, the mourning parents may serve as reminders of the social dislocation produced by war.

The Fox is certainly not exclusively about survival and fitness, as the massive amount of criticism on the story from countless critical perspectives clearly attests. However, Lawrence does indicate from the beginning of the novella that the struggle for survival is to be a central concern. There is a real ruthlessness to the way Lawrence writes of the deaths of even quite attractive but fatally weak characters in his fiction at this time, characters such as Egbert of "England, My England." In *The Fox*, Banford and March are trying to make a living by keeping poultry and a few cows: "Unfortunately things did not turn out well" (*Fox* 7). Not only is the farming enterprise itself on the edge of survivability, but Banford is repeatedly described as "delicate" in health. In fact, her father has given her the farm or the financial means to invest in it "for her health's sake, and because he loved her, and because it did not look as if she would marry" (*Fox* 7). At the age of thirty, it already looks as though Banford is a loser in the competition for survival and mating.

Bailey Farm seems at times to function like the Old Bailey: it proves to be the site for a rigorous trial of survivability. By the opening of the story, it has already, as it were, done away with two members of the older generation whom we see only in flashback: Banford's grandfather dies within a year of the beginning of the venture, and Henry Grenfel's grandfather has succumbed after the flight of his grandson to enlist (*Fox* 14). The two women are surviving at the opening of the story – "'We can look after ourselves,'" claims March (*Fox* 19) – but they are particularly inept with the animals when it comes to husbandry: "[J]ust before the other young beast was expecting her first calf, the old man died, and the girls, afraid of the coming event, sold her in a panic" (*Fox* 7). Even the fowls refuse to lay eggs. In effect, Lawrence wants to suggest that husbandry, the human imitation of the force of natural selection as Darwin maintained,[10] requires a husband. And so, after making the observation that "'there wants a man about the place,'" (*Fox* 17), Henry Grenfel suggests that he become just that with respect to March. He would marry her.

Lawrence makes it clear that Henry Grenfel, just back from Salonika with his military unit, is fit in physical terms. In contrast to the delicate and short-sighted Banford, Henry's "eyes were almost unnaturally clear and bright, though it was the brightness of abundant health" (*Fox* 15). Although his manners may leave something to be desired, at least as far as Banford is concerned, his senses are all keenly alive, and it is those five senses along with a sixth sense and an unusually strong will that allow Henry to kill the persistently plundering fox. In contrast to the earlier short-story version of the tale, in which the competition between Henry and Banford is a straightforward conflict between a vital male and a non-vital female for the love of a second woman, in the novella version, the sexual competition becomes conflated with a Darwinian struggle for survival itself. Having survived the war and clearly fit for survival himself, Henry becomes an agent of natural selection as well as sexual selection on Bailey Farm. In one of the earliest scenes dramatizing his enmity toward Banford, we see Henry standing under a wild pear tree "with the old dead leaves round his feet" (*Fox* 45), an obvious image of renewal through death, wishing the frail, struggling figure of Banford ill.[11] Lawrence uses the metaphor of the hunt to describe the rivalry between Henry and Banford for March's love but also initially for Henry's own conception of how he will have to go about accomplishing his desire to have March marry him (*Fox* 24). In the hunt, either the quarry escapes and the hunter goes hungry and perhaps dies, or the prey is killed and the hunter and his society feast. It is a battle for survival that has winners and losers.[12] By the end, in Henry's eyes at least, it is his life or Banford's that will prevail as in the struggle between hunter and hunted for survival: "The inner necessity of his life was fulfilling itself, it was he who was to live" (*Fox* 65–6).[13]

In *The Fox*, we might relate two types of death to a contrast encoded in the text between primitivism and civilization.[14] In the Neolithic Revolution, which took humankind from the hunter-gatherer form of economy to settled agriculture, there was, Joseph Campbell has argued, a drastic change in the male attitude to women. Women had been associated with death alone, but since death now included the mystery of a rebirth based on the seasonal cycle and vegetable recurrence, women came to be associated with the earth and rebirth as well, if we are to believe Campbell (129). There are generally two ways of dying in Lawrence. One leads to rebirth and is a symbolic death. The other is a physical death that is often both a thematic statement and an enabling part of Lawrence's narrative strategy. Lawrence would like to present Henry's arrival as an opportunity for March to be reborn through a painful process of dying out of the old way of love. Thus when Henry proposes to March initially, "[t]he world seemed to rock in her as if she were dying" (*Fox* 25). But it is Henry's change that is crucial for our purposes here. Henry's development suggests on a symbolic level a failed attempt almost literally to beat his sword into a ploughshare, to transform himself from a soldier/hunter to a settled farmer, a recapitulation in miniature of the Neolithic Revolution. The failure involves an ironic attempt to exorcise from his own being a predatory nature that has, presumably, been encouraged by the sanctioned killing during the war. Since the only way Henry can conceive of such a development is through the predatory killing of the totemic animal (whose death becomes a foreshadowing analogue to the later killing of Banford), the attempt to abandon primitive frames of reference serves instead to enhance them. It is by no means certain that Lawrence had such an irony in mind; in fact, from all appearances, Henry is meant to serve as a vitalistic hero. More than that, the primitive has reference here to ontogenetic development as well, so that the failure to abandon his killing ways also suggests Henry's inability, despite all the narrator's words to the contrary, to grow up and grow out of his dependence upon the mother figure, embodied at the end of the tale in his wife. That conscious intention and largely unconscious compulsions are working at cross purposes here is indicated, among much else, by the difficulty Lawrence had with the ending of the story and by its nicely balanced openness.

Although food is the most important of the actuating elements of the survival theme in the novella, confinement is significant as well. It is particularly with reference to Henry that we see the pattern of confinement and thwarted growth leading to a transformation that in this case proves deadly. Throughout *The Fox*, Lawrence plays a sense of confinement off against an ideal of natural freedom or expansiveness. The women's enterprise depends upon confinement, and part of their failure is an inability to impose limits: "One heifer, unfortunately, refused absolutely to stay in the Bailey Farm closes. No matter how March made up the fences, the heifer was out, wild in the woods, or trespassing on the neighbouring pasture" (*Fox* 7). This would be the case with any farm, but the fact remains that in *The Fox* Lawrence associates confinement with women. Henry, as a very young man, also refuses "to stay in the Bailey Farm closes," and runs away from his grandfather. But since this fact is "offstage" and past history it fails to figure significantly in the portrayal of confinement in the story. Henry's desire to develop, to "grow," to adapt, by becoming a husband and a farmer, is thwarted by Banford. The result is that the identification with the fox, which at first exists only in March's mind and, indeed, surprises Henry, becomes a reality. He *becomes* the very creature he kills; in fact, he can kill the fox only because of his own vulpine qualities. The fox becomes Henry's totem.[15] He becomes an animal on the prowl when he overhears Banford denouncing him to March: "His eyes were so round and wide that he seemed to see the whole night, and his ears were almost jumping off his head" (*Fox* 38). He walks "stealthily" away from the farmhouse and "prowled round the edge of the gorse cover" with "dilated eyes that seemed to be able to grow black and full of sight in the dark, like a cat's" (*Fox* 38). Like a fox hunting for a meal, "[h]e felt he could smell the hot, sickly, rich smell of live chickens on the cold air" (*Fox* 39). Nowhere else does Lawrence so perfectly justify Katherine Mansfield's complaint that Lawrence reduces people to "animals on the prowl" (Nehls ii 51–2).[16] Norris, surely, would argue that this is exactly what he wants to do as part of his biocentric agenda.

The text makes it quite clear, then, that even though Henry's killing of the fox is a symbolic attempt to exorcise from his own being the undesirable characteristics he shares with the animal, that attempt is ironic in its outcome. When he hunts down the fox, Henry for the first time feels a kinship with the animal. Realizing the fox will be trapped, he feels constricted himself: "And suddenly it seemed to him England was little and tight, he felt the landscape was constricted even in the dark" (*Fox* 54). Like the fox, Henry is an outsider posing his individuality and freedom of movement against the monotony and limits of civilized life that he senses in this scene in the form of fences, hedges, and domesticated dogs. The implications of Lawrence's choice of metaphors make it clear that it is, ironically, because of, not in spite of, his identification with the fox that Henry kills the creature. Henry's vulpine transformation bodes ill since the fox's success as a plunderer on Bailey Farm is offset by the threat of extinction hanging over his head (it seems to Henry that the fox "would be the last of the foxes" [*Fox* 39]), just as Henry's success in eliminating Banford as his rival for March is mitigated by the latter's unhappiness at the end of the story. If we express the development in Jungian terms for a moment – as Lawrence invites us to do by describing Henry's first perception of the fox as a shadow (*Fox* 39) – it would be fair to say that Henry fails because, although he may recognize his shadow when he sees it, instead of attempting to integrate it within himself, he tries to destroy it and, in so doing, destroys an aspect of himself that is necessary in a world that puts a premium on survivability. His triumph over Jill Banford, then, is ultimately ironic.

The Pyrrhic nature of Henry's victory is underscored by his use of a dead tree as his tool for the elimination of Banford. His arrival just in time to fell the tree represents the male claiming his prerogative as "husband" in two senses of the word and as leader. The natural world needs culling, and he will perform the task and at the same time eliminate from the human world the deadened and frail figure of Banford. In that sense, there is a strong identification between the dead tree and the unquick Banford, as F.R. Leavis claims (*D.H. Lawrence* 264n). The repetition of "weak" in the following passage underscores that connection: "Henry looked up at her, and met her queer, round-pupilled, *weak* eyes staring behind her spectacles. He was perfectly still. He looked away, up at the *weak*, leaning tree" (*Fox* 64, emphasis added). The tree in *The Fox* is a Scotch fir, but Lawrence's comments on the oak tree in the essay "Aristocracy" may nevertheless be applied here to suggest that the tree-felling is part of Lawrence's attack on traditional concepts of class hierarchy: "And the oak-tree, the slow great oak-tree, isn't he alive? Doesn't he live where you don't live, with a vast silence you shall never, never penetrate, though you chop him into kindling shred from shred? … You should seek a living relation with him" (*PII* 480). Where the women in *The Fox* merely wish to use the tree for firewood, Henry gets in touch with the tree's vital centre of power, establishing a "living relation with him," and in so doing demonstrating his own vitality, a vitality, we must recall, that allows him to kill his antagonist.

However, the dead tree, among much else, is also representative of Henry's own arrested growth; its felling by his own hands suggests that Henry must kill something in himself in order to destroy his adversary. Indeed, earlier in the story Lawrence describes Henry in arborial terms: "And he sat silent, unconscious, with all the blood burning in all his veins, like fire in all the branches and twigs of him" (*Fox* 55). Furthermore, the obvious phallic associations of the tree and Lawrence's use of the male pronoun in referring to it suggest a connection between the tree and Henry, whose influence March has been hacking away at since his departure, just as she has been chopping steadily away at the tree. So, although Lawrence would like to portray Henry victorious at the end, he is too honest a writer to shirk the implications of his own imaginative explorations. The boy is left at the end longing for a mythic Canada "with pain in his voice" (*Fox* 71).

March's transformation, a much more obvious one than Henry's, proves problematic as well. While most readers of *The Fox* continue to view March as a changed woman by the end of the tale, whatever their attitude to that change might be, the fact is that close examination of the key description of March's transformation reveals that it is a change largely in the mind of her lover and may well be nothing more than a perception coloured by his own wishes. The day before his leave is up and he is due back at military camp, Henry is surprised by March's unwonted wearing of a dress. To him, it transforms March completely: "No, she was another being. She was something quite different," and Henry feels changed himself by a new "grave weight of responsibility" and "the heaviness of male destiny" (*Fox* 49). Understandably, March can hardly fathom what Henry is making such a fuss over. "'What else do you expect me to wear, but a dress?'" she asks with some point (*Fox* 48). The end of the novella strongly suggests that March has not been transformed into the object of Henry's wishes and dreams and that they will have a difficult time of it in their marriage. All Henry succeeds in doing, really, is transferring his feeling of confinement onto March: "The moment she saw his glowing red face it was all over with her. She was as helpless as if she had been bound" (*Fox* 62). Not only has Henry's killing of Banford been self-destructive, it has also killed something in March instead of liberating her. The fatal wound in the fox, the "tiny streak of blood from one dark place" in the animal's fur that March touches (*Fox* 42) and the death wound at the back of Banford's head, "a mass of blood, of horror" (*Fox* 65), coalesce in the metaphor describing March's lethargy toward the end of the novella: "Something was missing. Instead of her soul swaying with new life, it seemed to droop, to bleed, as if it were wounded" (*Fox* 67).

Why does Henry not succeed? I would suggest that the failure is due to his compulsion, matching that of his creator, to insist upon inverting the parent-child hierarchy even as he thinks his greatest desire (which simply occurs to him one day out of the blue) is to marry March and settle down. Henry does succeed finally in rendering March childlike. When Jill is killed, March is like "a child that doesn't want to cry, but which is beaten from within, and gives that little first shudder of sobbing which is not yet weeping, dry and fearful" (*Fox* 66). We are a long way from the magnificent description in *The Rainbow,* the set piece of little Anna sobbing in the arms of her stepfather, Tom Brangwen, when her mother is in labour. Anna fights against sleep because she has not been allowed to see her mother. In *The Fox*, March, a grown woman, is described as fighting a sleep that Lawrence wishes to portray as the "natural" state not of a weary child but of womanhood in general: "And she was so tired, so tired, like a child that wants to go to sleep, but which fights against sleep as if sleep were death" (*Fox* 70). She even, finally, acquiesces in Jill's death in terms of survival of the fittest: "Jill would always be fretting herself thinner and thinner, weaker and weaker. Her pains grew worse instead of less. It would be so for ever. She was glad she was dead" (*Fox* 68). Her prophetic second dream had suggested an unconscious collusion with Henry's predatory relationship to Jill, for the language insists that she herself has become a hunter: "So she hunted and hunted, and picked up thing after thing, and threw it aside in the agony of dream-frustration" (*Fox* 40). What she is looking for is something with which to cover the dream-dead Banford. Her resorting finally to the fox skin is clearly meant to suggest an unconscious death wish toward Banford since Henry, by this point, has succeeded in shooting the fox.

It is ironic that one of the things that determines March for Henry instead of Banford in the first place may be the promise of producing a child together. At the end of the proposal scene, March helps Henry bring in the logs he has been sawing for firewood, "filling her arms and carrying the logs on her breast as if they were some heavy child" (*Fox* 27). It may not be too fanciful to suggest that, in the scene where March finally agrees to marry Henry before he leaves for Canada, the terrible signalling "from beyond" that March senses in her suitor's heartbeat and that renders her "paralysed" and "helpless" (*Fox* 52) may represent the heartbeat of the child she longs for and wishes to internalize in a literal sense.

Jane Nelson has suggested that the "gaze behavior" between the fox and March at the beginning of "Lawrence's murderous little tale" suggests that between an infant and its mother and that "such gaze behavior may play a role in evolution, since the infant's success in capturing and holding his mother's gaze helps assure his biological survival" (138). As the substitute initially for the gazing fox in March's mind, then, Henry suggests a child until he insists on his fitness as a mate for March and thus the potential co-producer of a child. However, Henry's killing of Banford by tree-felling may indicate the sterility of his future relationship with March. At the end of the tale, Lawrence tells us that, although Henry "possessed her in every possible way" and that March "wanted nothing else, now," sexual fulfillment seems not to be enough; "[s]omething was missing" (*Fox* 67). This "something missing," I am suggesting, may be a child. For, instead of producing a child with March, Henry turns her into one. Nevertheless, he has succeeded in his covert quest, the child's quest for mastery over the mother to ensure his survival. He wanted so much "to make sure of [March's] soft woman's breasts under her tunic" (*Fox* 47) because he needs reassurance that March will be a source of nourishment, of life, to him.

Finally, the failure at the end of the story is a mark of Lawrence's doomed narrative strategy combined with his integrity as a writer. His inability to understand his own compulsion for control in male-female relations, derived from his experience of an unusually intense and dependent childhood bond with his mother, throws a spanner into the works of his strategy to link the unnatural life of a liberated woman to the need in society for male leadership. It is an effort and a result that will be repeated with fascinating variations in *The Captain's Doll* and *The Ladybird*.

THE CAPTAIN'S DOLL: DEFENESTRATING THE PROBLEM

The Captain's Doll tells of Alexander Hepburn, a forty-ish captain whose hobby is astronomy, and of his affair in postwar Germany with a poverty-stricken aristocrat, Johanna zu Rassentlow (familiarly known as Hannele), while away from his wife, Evangeline, and their children in England. In order to survive, Hannele, together with her friend Mitchka, creates and sells dolls and other handicrafts, and at the opening of the story she is fashioning a doll that is a faithful copy of her lover. Hepburn's wife arrives in town determined to put an end to the affair, accounts of which have reached her in England. However, in a sudden twist in the plot introduced by the words "And then a dreadful thing happened: really a very dreadful thing" (*Fox* 109), Mrs Hepburn "obligingly falls victim to vertigo" (Harris 159) and falls out of her hotel window to her death, leaving her husband financially independent. Shortly thereafter, Hannele journeys to Austria, where she becomes engaged to a government official, the Herr Regierungsrat von Poldi. Locating her there with some difficulty, Hepburn invites her on an excursion to the Tyrol Mountains, where, after his successful climb of a glacier, the two of them debate the terms of love and marriage. In the end, Hannele agrees to honour and obey her future husband as provided for in the Church of England marriage service. While the obvious image pattern of doll, moon, and glacier carries the burden of the plot, the less apparent configuration created by images of power, sexuality, and food bespeaks a subliminal design that points to the influence of evolutionary concepts. Even as Lawrence's male protagonist engages in a process of proving his fitness, even as both central characters are involved in sexual selection, the author himself resorts to a ruthless weeding out among his characters in order to bring about the desired denouement. As in *The Fox*, the ending leaves the reader feeling justifiably puzzled. But the trouble is not only with the end of the story. There is a basic incongruity in *The Captain's Doll* that reflects the ongoing confusion created by the clash of Lawrence's conscious narrative strategy and his largely unconscious pre-oedipal compulsions.

We saw how, in *The Fox*, the description of March touching the dead fox has an oddness to it that suggests a deeper inconsistency in the novella that works against the author's narrative strategy. In *The Captain's Doll*, the most blatant element of incongruity involves Hepburn's description of his wife to Hannele, after his wife has fallen to her death. She was a "clever little monkey" evidently unable to advance beyond the simian stage of development. He explains to Hannele that Evangeline is better out of the world since "she was born in the wrong period – or on the wrong planet" (*Fox* 110). Unlike Thomas Hardy's Tess, another figure born in the wrong era,[17] it is not the power of the state that eliminates Mrs Hepburn but a kind of cosmic nemesis that weeds out those slow to adapt, a force that seems very similar to Darwin's principle of natural selection but which, in the last instance, is wielded by the author himself. The trouble is that the Evangeline Hepburn Lawrence has portrayed with deft strokes in the novella in no way matches the woman described by her husband in this scene, even though he is apparently sincere and, given the narrative strategy at work in the story, he seems to have authorial backing. The picture Hepburn paints of a helpless, otherworldy creature is simply not credible. It is so far removed from the active, authoritative, and decisive woman we have seen that a more convincing interpretation for the scene is in order than those that have been proposed by the story's critics to this point. F.R. Leavis was right to say of Hepburn's strange description of his wife that "it is the reality, not of the wife thus presented to us, but of Hepburn himself on which our wonderment focusses during the astonishing presentment." It seems to me, however, that his conclusion that the main effect of the scene is to suggest "the unknown in Hepburn" is inadequate and even misleading (*D.H. Lawrence* 212). Later critics have fared little better with this puzzling portion of the novella.[18]

To Hannele's and the reader's amazement, Hepburn compares his wife to a pet black-cap he had caught and caged as a boy. The bird is dependent upon the young Hepburn for its food as the older Hepburn is upon his wife for his tea: "'And it would peck its seed as if it didn't quite know what else to do; and look round about and begin to sing.'" But it soon dies. The comparison of Mrs Hepburn to a bird – she too "'would pick her seed and look round in the cage'" (*Fox* 112) – begins to make some sense within the context of a Darwinian struggle when we reflect that Lawrence repeatedly likens his hero to a cat or a "great cat" (*Fox* 92). In "The Reality of Peace" (1917), Lawrence wrote, "we must eat life to live. We must, like the leopard, drink up the lesser life to bring forth our greater" (*P* 681). It is also interesting that Hepburn initially is both tomcat and kitten for Hannele, so that the confusion between mate and parent that characterizes his marriage seems to be repeating itself early on in the new relationship. The scene's otherwise anomalous presence is comprehensible, then, if we view it in terms of Lawrence's application of evolutionary concepts in his narrative strategy at a point in his career where male leadership increasingly becomes a vehicle for misogyny. It also reveals Hepburn's capacity for self-delusion.

Hepburn is not at all the deferential, self-abnegating lover he thinks he is early in the story: "'Why bless you, I'm not important a bit,'" he insists to Hannele (*Fox* 82). His self-portrait is as distorted as the portrait he paints of his wife and deserves the corrective effected by Hannele's doll. In reality, he is self-centred and insecure because of unfulfilled needs. His protest on the way to the glacier that he is greater than the surrounding mountains, something Hannele views as "megalomania" (*Fox* 138), reflects an attempt by a vulnerable self to restore the infantile feeling of omnipotence. This aspect of Hepburn is pushed underground, as it were, only to emerge in projective imagery. We cannot understand the story without understanding Hepburn, and we cannot understand Hepburn if we do not see the connection between his personality and the imagery of the novella. Hepburn's sense of constriction, for instance, which he has attempted to lessen by means of his astronomical hobby,[19] is projected and inverted through his comparing his wife to various confined pets: "[S]he was always arranging herself and fluttering and chattering inside a cage" (*Fox* 111). In the present situation, Hepburn would have liked his wife to have remained "caged" at home in England so that he might indulge in his love affair. Instead she arrives unexpectedly to constrict his freedom. As Jill Banford controls the coveted farm in *The Fox*, Mrs Hepburn has hold of the family pursestrings in *The Captain's Doll*. Since he is economically dependent on her, when she arrives in Germany, Hepburn does not dare force the issue in spite of prodding from Hannele. Furthermore, it is only after his wife's death that Hepburn frees his daughter from the convent she has been living in (*Fox* 114). Despite his protestations to the contrary, Hepburn has considered his wife more a jailer than a prisoner.

There is a struggle for survival in *The Captain's Doll* based on an imbalance of power in a historically determined situation. We must remember that following the First World War parts of Germany were occupied by British troops (among others) whose function it was to execute the humiliating terms of the Treaty of Versailles. As his letters of the time suggest, Lawrence was acutely aware of the situation in the countries defeated in the war.[20] Mitchka fears Captain Hepburn, and her anxiety may be related to the fact that he has power as part of a military occupation. The doll, says Mitchka to her friend, "'makes me afraid. And you have no right to get me into trouble do you see. It is not I who look at the English officers'" (*Fox* 86).[21] For Hannele, however, sleeping with the enemy seems to be a good tactical move in the struggle for survival. On the other hand, a false step can be dangerous: "for this dead puppet she had been compromising herself, now, when it was so risky for her to be compromised" (*Fox* 90).

In Germany, Mrs Hepburn is quickly able to determine where the lines of power run and how to make judicious use of them in her attempts to frustrate her husband's desire to discard her in the game of sexual selection. Upon first meeting Hannele and Mitchka, she subtly insinuates the idea of expulsion into the conversation by quoting Goethe: "'it means you always feel you would be happy somewhere else, and not just where you are'" (*Fox* 88). When her hint fails, she makes it clear to Hannele that she, rather than the captain, controls the family income, for he "'has no income of his own'" (*Fox* 98, 102), and that she can easily have Hannele and Mitchka removed from the city by the British authorities if necessary. She can also, she claims, get her husband "a post in one of the observatories" if she can persuade him to resign his commission and return to England (*Fox* 103). She is obviously a woman with connections. In a letter to Hannele, she writes, "Our old family friend, Major General Barlow, called on me yesterday … It seems we have full power to send away any person or persons deemed undesirable, with twenty-four hours' notice to leave" (*Fox* 109). The account of Mrs Hepburn's death follows immediately upon the description of Hepburn reading this same letter (which, ironically, renders the writer herself "undesirable") just as in

The Fox Henry returns to the farm to do away with Banford after receiving a letter from March. Her "dire threat" (*Fox* 103), in any case, does not reflect power of the vitalistic sort that Lawrence wants to ascribe to Hepburn but rather willful bullying.

Caught between the aphrodisiac of power in the captain and its threat in his wife, Hannele, nevertheless, has a power all her own, that of sexual selection. In fact, as in the story "The Blind Man," all three novellas in the *Ladybird* volume in different ways demonstrate how ostensibly powerless individuals gain ascendancy in vitalist terms. Although the theme of vitalistic ascent is associated in this novella primarily with Hepburn rather than with Hannele,[22] she has the power, apparently, to induce Hepburn to throw over his wife and children for a life with her. Her power becomes a major issue in the rather farcical debate at the end of the novella, but for most of the story, Hannele's physical attraction to Hepburn leaves her "powerless" (*Fox* 84).

The issue of power is, then, closely related in *The Captain's Doll* to that of sexuality. Hannele is quite right in her belief that Hepburn "wanted to make her love him so that he had power over her. He wanted to bully her, physically, sexually, and from the inside" (*Fox* 139). We ought to bear in mind that the title of the tale refers not only to the doll that Hannele makes of the captain but also to Hannele's position in the relationship. She is the captain's doll in the sexual meaning of the word, which the OED gives as "a pet form of the name Dorothy. Hence given generically to a female pet, a mistress."[23] The emphasis that Hannele places on the doll's sexual representation – Mitchka asks, "'Has he really such beautiful fine legs?'" (*Fox* 76) – is analogous to the inevitable centrality of sexuality in a vulnerable woman's relationship with a man who is married and has two children. That Hannele also needs the doll to remind her, when Hepburn is absent, of his mysterious attraction suggests that the puppet is also an implicit protest against her role as a mere adjunct to Hepburn's otherwise full life, as her bitter complaints that she matters so little to him also imply (*Fox* 81–2). Hepburn's determination to retain control in his relationship with Hannele is a function of the pre-oedipal patterns that, as in *The Fox*, emerge primarily though covertly through imagery of food and eating.

In a letter dated a few months before the composition of *The Captain's Doll*, Lawrence wrote to Mary Cannan: "I would like to break out of Europe. It has been like a bad meal of various courses – Europe – and one has got indigestion from every course" (*L* iv 49). Lawrence's view of the importance of food in this period was demonstrably linked, unconsciously no doubt, to the concepts of struggle and hierarchy, as the previously quoted passage from "Reflections on the Death of a Porcupine" suggests (see page 17 above). In *The Captain's Doll,* the fight for food represents a more general struggle for survival, a struggle that Lawrence makes sure is won by his representative natural aristocrat.

The presence of food and eating in *The Captain's Doll* is unobtrusive because Lawrence is so very good at rendering such mundane aspects of daily life, a legacy from the realism he perfected in *Sons and Lovers.* Once one begins to note the frequency and location of the motif, however, one is struck by its pervasiveness. Tea parties constitute a veritable leitmotif in *The Captain's Doll,* while the need for oral gratification all round seems insatiable. In defeated Germany, the struggle for survival includes the strain of obtaining sufficient food. Hannele and Mitchka, in fact, are living a hand-to-mouth existence. No matter what else he may be to Hannele, the captain also represents a meal ticket. Her exasperation at the opening of the novella at his failure to appear is related to food: "It was long after dinner-time – why hadn't he come?" (*Fox* 78). Is Hannele more impatient to see her lover or to have a square meal? When she discovers that Hepburn will be joining his wife at her hotel, Hannele muses on such an attachment in terms of cuisine: "'I suppose,' she said to herself, 'that is love à l'anglaise. But it's more than I can swallow'" (*Fox* 95). Strangely, though the heroine here seems to be choking on a gnat, so to speak, by the end of the story she will be willing to swallow an elephant.

Hannele's interest in food is justifiable enough. Only her artistic abilities and those of Mitchka ensure that "the two women did not starve" (*Fox* 85). It is striking that, in Hepburn's absence from town early in the story, Hannele wonders about the reality of the man in decidedly gustatory terms: "But this English officer he was neither fish, flesh, fowl, or good red herring, as they say" (*Fox* 90). When his charm ceases to work on Hannele, she "tasted salt[;] … distaste was in her mouth" (*Fox* 106, 107). In Austria, Hannele's chances of survival improve with her attachment to the Herr Regierungsrat von Poldi, but the situation in general is no better since the tourists who swarm to Kaprun in the summer "ate up the food like locusts" (*Fox* 121). Austria is in the throes of a craze for physical fitness – "Such magnificent naked limbs and naked bodies" (*Fox* 122) – that is meant to be contrasted by Hepburn's vitalistic sort of fitness, despite the numerous descriptions of his physical attractiveness. Hepburn is able to trace Hannele to Kaprun through the "Chit-Chat column of the *Muenchener Neue Zeitung*" (*Fox* 118), which mentions a modernist still-life painted by Theodor Worpswede that includes the doll Hannele made of the captain. The column goes on to indicate Hannele's current whereabouts in Austria, but more to the point is the description of the doll surrounded by objects related to food: "two sun-flowers in a glass jar, and a poached egg on toast" (*Fox* 118). It is as if the doll in this novella had its origins in the Greek fairy tale "Mr Simigdáli or the Gentleman Made of Groats," which tells of a princess who makes a man for herself by mixing almonds, sugar, and groats and infusing him with life through prayer (Lüthi 87). What Max Lüthi has said of the Greek princess could apply equally to the impression Lawrence is attempting to create with reference to Hannele and her creation of the doll: "She does not love another human being, she loves an image that she has made for herself ... which ... is sweet, and she gives herself up to it the way Hansel and Gretel do to the sugar in the gingerbread house ... She does not really love the other person, she loves the sweet image which originated within her" (92). The trouble is that Hepburn is not sure if Hannele is a Gretel or a witch. In *The Captain's Doll*, love as consumption frightens off a middle-aged man who has had to cope with a devouring mother all his married life.

The imagery of food – the basis of individual survival – becomes concentrated during the excursion to the glacier, as Hepburn implicitly argues through word and action that he, rather than the Herr Regierungsrat, offers Hannele the better chance of survival under the new, postwar dispensation. Lawrence, for example, lovingly describes the "long, hot loaf of pure white bread, beautiful sweet bread" that Hepburn has bought for the climb (*Fox* 127). While the unmistakable sexual connotations embedded in the description of the mountain valley have often been remarked upon,[24] the food and eating metaphors and the activity of eating itself have been less noticed. In this Tyrolean valley full of raspberries, bilberries, blackberries, and cranberries, the water is an "insatiable" force sadistically "biting in into the rocks' intensity ... like a beast of prey" (*Fox* 129), while up on the slopes there are feeding cattle (*Fox* 132). The two excursionists stop several times to eat or drink, and see others doing the same. Lunch, we learn, consists of bread, wine, cheese, sausages, and eggs; there is a later snack of hot coffee and milk at a restaurant where the "fair youths" boast "huge *voracious* boots" (*Fox* 140, emphasis added). When Hepburn wishes to damn the mountaineers around them, his diction is telltale: "'I'd like to make them all stop up there, on their mountain-tops, and chew ice to fill their stomachs'" (*Fox* 137). Hepburn finds the glacier hateful apparently because it seems poisonous to him (*Fox* 140) and because he prefers "'the world where cabbages will grow on the soil ... [T]o me the ice didn't look good to eat, even for a flea'" (*Fox* 144). Even an upholstered chair on a wagon wheeling by is gold and orange "like fruit juice" (*Fox* 133). One can almost imagine Hepburn salivating, for he is a greedy man, as his wife once told Hannele. Mrs Hepburn emphasizes her husband's almost compulsive attachment to his tea when the two women meet in the Domplatz and the Englishwoman invites the German up to her hotel: "'My husband *adores* his tea. I don't suppose our man is five minutes off the half-past, ringing the gong for tea, not once in twelve months ... [H]e gets – quite – well, quite "ratty" if tea is late'" (*Fox* 95). Once in the hotel room, Mrs Hepburn assures her guest in Hepburn's presence that "'[m]y husband is a greedy man, a greedy man – he likes at least three cups – and four if he can get them – or *five*!'" (*Fox* 96). True to form, the captain swallows the five cups of tea his wife allows him before trotting off on a transparently arranged errand.

The emphasis on food is perhaps appropriate for a man who plans to travel to Africa to help get "three thousand acres of land under control" (*Fox* 152). Moreover, the prospect of an abundant supply of food in Africa seems to eliminate Hannele's uncertainty regarding the prospect of marriage with Hepburn. He assures her that he will not merely "sup it up" like a Sultan as she puts it (*Fox* 151) but engage in farming and write a book about the moon. In this he presumably demonstrates a Lawrentian balance between the needs of the body and those of the mind or spirit. But the moon has been indirectly responsible for his wife's fatal fall, and even farming is not a purely creative activity. "But you have to fight even to plant seed," writes Lawrence in "On Human Destiny." "To plant seed you've got to kill a great deal of weeds and break much ground" (*PII* 629).

The food/eating motif takes the story into the orbit of Lawrence's own struggle with unresolved pre-oedipal problems. Not only does Hepburn resent the loss of freedom that his wife's arrival signals, but, it could be argued, he finally recognizes the fact that he has played the role in his marriage of the dependent and obedient child. Seeing his wife's prematurely aged face after a long absence and hearing Hannele's reaction make that plain to him. When Hepburn tells Hannele that the little, elderly lady who was so anxious to purchase her Alexander doll is his wife, "she looked up really astonished. She had thought it might be an acquaintance – perhaps his aunt – or even an elder sister. – 'But she's years older than you,' she added" (*Fox* 93). In fact, the difference in age between the Hepburns is slightly less than that between Henry and March in *The Fox*. In both stories, the woman is about ten years older than the man (Frieda was six years older than Lawrence), but in *The Captain's Doll*, the male protagonist will attempt to invert the mother-child hierarchy by escaping the mother figure altogether. In that effort, Mrs Hepburn's "wrinkled" and oddly "crumpled" face (*Fox* 86) serves the same function in *The Captain's Doll* as Banford's weak eyes do in *The Fox*. It is a mark of her unfitness.

The parent-child paradigm is in Hepburn's mind when he describes his wife to Hannele as being "like a child with playthings" (*Fox* 112) and her eyes as being "[l]ike a child's, that is listening to something, and is going to be frightened" (*Fox* 111–12). He has simply inverted reality. In actual fact, suggests the narrator (in terms that might describe a toddler who has lost its soother), there has been "a long slow weaning away" for Hepburn, while his wife's death is a "sudden silent rupture" (*Fox* 114). In *The Captain's Doll*, the need for a mother is both acknowledged and bitterly fought. In fact the term "weaning away" calls to mind Lawrence's description of his own difficulties in gaining independence from his mother: "I had a devil of a time getting a bit weaned from my mother, at the age of 22" (*L* i 527).

The redefinition of love in this novella is inextricably connected to the pre-oedipal pattern. At the end of the story, Hepburn believes that he no longer wants love. "'Whom have you ever loved?'" asks Hannele with some pertinence. "'To begin with, my mother: and that was a mistake'" is the reply (*Fox* 149). It begins and ends with the mother, but Lawrence, at this stage in his development, can break his protagonist of his dependence only by imagining both a process of evolution within the individual that ruthlessly expunges such a need and a corresponding outer force that just as harshly eliminates its object. Mrs Hepburn has been blocking her husband's vitalistic and relational growth, and she is removed. Von Poldi, Hepburn's rival for Hannele, however, is simply jilted. Both are portrayed as hopelessly mired in a past world, unable to navigate the treacherous waters of the onward flow of vitalistic development. In "Education of the People," Lawrence wrote that the only way to save the sanity of our children is to pull down the mothers "from their exalted perches" (*P* 621). It is appropriate, then, that the mother figure, Evangeline Hepburn, falls to her death while the husband-child who once worshipped at her feet successfully climbs a glacier as his new woman watches from below. Hepburn's objection to the doll, from this perspective, is clearly related to the traditional function of dolls as substitute children in the maternal role playing of young girls.

As in *The Fox,* Lawrence uses class as a motif to underscore his redefinition of hierarchy. Lieutenant Friedeburg, Hepburn's predecessor in the novella's ancestral tale, "The Mortal Coil," is an aristocrat with all the values of his class. Marta, the early Hannele figure, looks "at his aristocratic fineness of physique" (*PII* 70). By way of contrast, Hepburn is in social terms middle class, as is his wife. But where the captain seems neither terribly impressed by nor dismissive of aristocratic titles, Lawrence makes Mrs Hepburn what Leavis calls "classily vulgar" (*Thought* 94). She professes to feel sorry for Hannele and Mitchka: "'It must be very hard to have to do these things for a living, after you've been accustomed to be made much of for your own sake and for your aristocratic title.'" She views the two women as déclassé aristocrats: "'Better drop the titles altogether, I think …,'" she says to her husband (*Fox* 91), shortly after her own overly demonstrative and falsely obsequious use of the titles in addressing the two women in their studio. Mrs Hepburn herself has class pretensions that are exposed in her letter to Hannele, which is written in "her well-known, large, 'aristocratic' hand" (*Fox* 108). Lawrence had very little time for such class consciousness. Perhaps the most dismissive words of all are reserved for what the text repeatedly and distastefully terms "Jews of the wrong sort." What "the wrong sort" means in this context, apparently, is a putting on of aristocratic airs "so that you might think they *were* Austrian aristocrats, if you weren't properly listening, or if you didn't look twice" (*Fox* 140). More generally, the narrator of *The Captain's Doll* warns us of the dangers of class hatred, which can "swallow all human feeling in its abyss" (*Fox* 128), a phrasing similar to the description in *The Fox* of the "illusion of attainable happiness" as a "void pit which can swallow you and all your efforts into its emptiness, and still be no emptier" (*Fox* 69). The ogre of the traditional class system is a terrifying devouring figure, and its association with Mrs Hepburn suggests the devouring mother. Hepburn does not so much see his wife as a pet bird eating inside a cage as he sees himself caged by his wife and in danger of being eaten.

As for Hannele, she is rather disgusted with Hepburn once she sees him in the domestic setting with his wife, acting "like a servant" (*Fox* 105). Now he seems "[a] limited, inferior, slightly pretentious individual!" (*Fox* 106). The Herr Regierungsrat, in contrast, may appear "at first glance lower middle-class," but, in fact, "he wasn't." This "ill-bred-seeming" man is "not ill-bred at all" (*Fox* 119). Hannele has yet to learn the lesson of Lawrentian hierarchy. Von Poldi may have a touch of the natural aristocrat about him – and this touch may be his salvation in terms of the story's narrative strategy – but he is associated with the decadent late emperors of Rome while Hepburn is a Caesar, the founder of the line. As in *The Ladybird*, Lawrence is arguing for the true power of a natural aristocracy (and thereby subsuming the hierarchy of class to that of survival). As Beer suggests, Darwin's evolutionary concepts were subversive of the class organization in nineteenth-century England (63). But Hepburn's social ascent from an alliance with middle-class money to a union with upper-class prestige – taking the money with him! – suggests the purest wishful thinking on Lawrence's part and gives Hepburn's climb of the glacier the added, if unintended, significance of a symbolic social climb.

Mrs Hepburn's death is "justified" by her unfitness, her inability to adapt, as dramatized in her poor judgment regarding the identity of her husband's mistress and elaborated in her crumpled face and in Hepburn's likening her to a monkey and a caged bird. Alexander Hepburn, on the other hand, is intended to be "fit" in the struggle for survival, and his climbing the glacier is meant to convince Hannele that he is the male moving purposefully forward in life. Hepburn's fitness for survival is set off both by the general decadence of Austrian society and by the Herr Regierungsrat, an "old Roman of the decadence" (*Fox* 120). In the terms of the Hardy essay, the Herr Regierungsrat, who "seemed as if he had no legs, save to sit with" (*Fox* 120), is a personification of the Will-to-Inertia, while Hepburn represents the Will-to-Motion, thus his fine legs. For Hannele, "[b]eing in Austria was like being on a wrecked ship, that *must* sink after a certain short length of time. And marrying the Herr Regierungsrat was like marrying the doomed captain of the doomed ship" (*Fox* 121). Her choice of captains is clear enough. Hepburn's climb also spells defeat for the forces of female devourment, for the glacier, with its "great tongue," its "great fangs and slashes of ice" that bite "into the flesh of the earth" (*Fox* 132, 133), is the ultimate symbol in *The Captain's Doll* of the threat of being eaten.

By the end of the story, Hepburn is a far cry from the man Mitchka could justifiably describe as "finished off" (*Fox* 76) and "at a closed end" (*Fox* 77). Hannele once compared his smile to "those Chinese carved soapstone apes" (*Fox* 82), and the captain himself to an ammonite, the fossilized remains of an extinct species (*Fox* 94). His arrested growth has resulted in a transformation that is healthier than the corresponding change in Henry Grenfel, but the cost in terms of female life is the same, and Lawrence does not seem to mind that he is patently fudging the issue much more than in *The Fox*. Once again, there are two kinds of death in the story: a physical death reserved for the meddling female and a symbolic death as a prelude to rebirth: "a man hasn't finished his life at forty. He may, however, have finished one great phase of his life" (*Fox* 115). There is a startlingly violent description of Hepburn after his wife's death that is quite similar to the description of Jill Banford's messy end in *The Fox*: "The chief thing that the captain knew, at this juncture, was that a hatchet had gone through the ligatures and veins that connected him with the people of his affection, and that he was left with the bleeding ends of all his vital human relationships" (*Fox* 113). The point, of course, is that the description of Hepburn is metaphoric while that of Banford is literal. Within *The Captain's Doll* itself, Hepburn's wife must die so that Hepburn may be reborn, and she *deserves* to die, the imagery tells us, because she is unfit. While, in *The Fox*, the fox dies as a kind of ritualistic prelude to Banford's death (Vickery 81), in *The Captain's Doll*, the doll itself and the still-life painting of it function as the thorn in Hepburn's side that Banford was to Henry (*Fox* 151, 59). Hepburn can be content with destroying the symbol of what he despises in Hannele (and presumably all women) by destroying the painting, but by the logic of the story, Hepburn himself – like his wife, compared to an ape – is quite unfit, unadaptable in the extreme. He simply has no idea what to do when Evangeline arrives in town. Yet Lawrence allows him his rebirth and transformation while denying the same to her. Hepburn, Lawrence tells us in terms of his characteristic vegetative metaphor, "began to bud with a new yearning for – for what?" (*Fox* 115).

At the climax of the novella, Hepburn is able to defeat the same ice world that in *Women in Love* killed Gerald Crich, another man who comes to his woman as a child to its mother.[25] Though Hepburn triumphs where Gerald fails, it is nevertheless ironic that his entire performance on the excursion is childish, not to say infantile. In climbing the glacier, he must go "ridiculously as on four legs" (*Fox* 142), and the mountain valley, "that first rocking cradle of early water" (*Fox* 133), has the homey feel of the nursery. So even though it appears that Mrs Hepburn fails to survive because she is an evolutionary throwback, Hepburn himself must, it seems, incorporate regressive tendencies on both the ontogenetic level (his orality) and on the phylogenetic level (going on all fours) before moving forward.[26] Lawrence would like to portray his hero as one of the people he describes in his foreword to *Women in Love* as able to "bring forth the new passion, the new idea" (*WL* 486) and therefore able to endure in the struggle for survival and the evolution of humanity, but his honesty once again prompts him to include material that suggests regression stemming from artistic, even moral, confusion. The irony inherent in Hepburn's climb suggests that Lawrence realizes the extremes to which his narrative strategy has taken him, but he persists to the end, nevertheless. This is perhaps the point in the story Lawrence refers to in one of his letters where he speaks of having taken his tale high up into the mountains and not knowing quite "how to get it down without breaking its neck" (*L* iv 109). In the end, it is the broken neck of Mrs Hepburn that allows Lawrence to complete his story in the way he wishes.

THE LADYBIRD: INTO "THE ARMS PLUTONIC"

If *The Fox* attempts to portray March as a Sleeping Beauty (although one who must be put *back* to sleep after her awakening), *The Captain's Doll* transfers the archetype, for all intents and purposes, to the male protagonist. Hepburn is far more in need of awakening than Hannele, although Lawrence turns the tables somewhat disingenuously in the story's second half by having Hepburn ride in as the knight in shining armour to "rescue" Hannele from the Herr Regierungsrat and the dragon of her own worst impulses. In both *The Fox* and *The Captain's Doll*, the character invested primarily with authorial identification seems to be the male protagonist. In *The Ladybird*, in a curious way, Lawrence seems to a great extent to identify with Lady Daphne, even though she is manifestly the character who is awakened into vitality in this novella rather than the Prince Charming who does the awakening or rescuing.[27]

The Ladybird deals with a wounded prisoner-of-war aristocrat from Central Europe, Count Dionys Psanek, who lies close to death in a British hospital for enemy wounded. He had known the Beveridge family before the war, and Lady Beveridge determines to visit him despite the loss of her own two sons in the war and the incarceration in enemy territory of Basil, her daughter Daphne's husband. Slowly, with the help of Lady Beveridge and Daphne, the Count regains his strength and health, and he himself awakens, in typical Lawrentian fashion, the suppressed passions of Lady Daphne, who has always been rather sickly but now becomes much more robust. At war's end, Basil returns home, and Dionys is invited to spend some time at Thoresway, the Beveridge ancestral estate, before he returns to his homeland. A strange sound in the night draws Daphne to the Count's room. It is his singing, but to Daphne it is a sort of summons from the beyond or from her own untapped unconscious. She allows herself to become "enfolded in the deeper dark / of the arms Plutonic" in the words of Lawrence's poem "Bavarian Gentians" (*Poems* 697), and they become lovers. When the Count leaves, he promises that he will always be there as her night husband in a kind of updated Pluto-Persephone myth.

Although the short stature of the Count would suggest that his growth has been curtailed in a literal way and that he is, according to Lawrentian formula, in need of transformation, it is Daphne who undergoes the greatest transformation or at least the most genuine one in the story.[28] Like the male figures of the two companion novellas, she is stifled by a maternal bond that is repeated in her choice of husband. (In fact, Daphne feels "suddenly stifled in her closed furs" when she first visits the Count in the company of her mother [*Fox* 165].) Vitally, both Lady Beveridge and Basil represent what, in Lawrence's estimation, is the outdated world view of Christian love. Daphne has "been brought up by her mother to admire only the good," and "her will was fixed in the determination that life should be gentle and good and benevolent" (*Fox* 161). The problem is she has her father's blood in her, the blood of "a daredevil border soldier" (*Fox* 161). The later Satanic associations of Dionys suggest that he is the daring devil she needs for her healing transformation to be effected. Like March in *The Fox*, however, her vitalistic awakening leads only to a sleep. In Daphne's case, it is a literal sleep, for, during her affair with Count Dionys, Daphne "had a strange feeling as if she had slipped off all her cares … She felt she could sleep, sleep, sleep – for ever" (*Fox* 217). Lawrence's version of Sleeping Beauty is, from the woman's point of view, rather circular: an awakening in vitalistic or sexual terms leads to a soporific state in metaphorical and/or literal terms. By way of contrast, the awakened male remains alert. Similarly, the "blocking" character in this novella is male rather than female, so that the details of the pattern we have seen in *The Fox* and *The Captain's Doll* are a little different. On the other hand, the portrayal of the evolutionary theme of fitness and extinction and the proposed solution of a revised social hierarchy are obvious in *The Ladybird*, as is the inhibiting influence of what we may continue to call pre-oedipal concerns.[29]

A key incongruous passage in *The Ladybird* comes towards the end of the novella, just prior to the presumed sexual union between Dionys and Daphne. Dionys's uncanny singing late at night while a guest at Thoresway, he explains to Daphne, involved a song of his country about a swan who became a woman because she wished to marry a human hunter. "Then in the night one night the king of the swans called to her to come back, or else he would die. So slowly she turned into a swan again, and slowly she opened her wide, wide wings, and left her husband and children" (*Fox* 215). I would like to reserve my discussion of the full implications of this fairy tale until later and focus here on the incongruity and what it may indicate.

It seems odd for Lawrence to insert at this point in his story the tale of a female who must negate a happy transformation she has undergone, for he is about to suggest that Daphne is going to be changed by Dionys's love forever. Furthermore, on the purely realistic plane of the story, it is Dionys who is about to depart for his home country, not Daphne. He, rather than her husband, Basil, is the king in vitalistic terms. However, the key to the passage, I believe, is the information that the king of swans will die without the ministrations of his female. If food was central to an understanding of the issue of survival in *The Fox* and *The Captain's Doll*, in *The Ladybird* it is illness and its cure.

At the opening of *The Ladybird*, we meet three characters who are, in one way or another, damaged in the chest at a time in history that, in Lady Beveridge's words, "'is like a great sickness – like a terrible pneumonia tearing the breast of the world'" (*Fox* 163). The "pierced heart" of the martyr-like Lady Beveridge has suffered the pain of countless swords, but like the void pit that gets "no emptier" in *The Fox*, there always seems room for another, and she, like the true Christian she is, will continue to love her enemies (*Fox* 157). Her wounds suggest the bosom as a merely metaphorical site of injury, while her daughter is threatened with consumption and Dionys has taken a bullet through the chest. Illness, like food, can represent both the struggle for survival and a dependency in the parent-child relationship. Certainly, in Lawrence's own case, his youthful illnesses became defining moments in his relationship with his mother even as his care for her in her final sickness in some way must have represented an inversion in that dependency that finally helped to liberate him as far as he was ever liberated. Where in *The Captain's Doll*, as we have seen, Hepburn and his wife engage in an antagonistic rivalry of feeding each other, in *The Ladybird*, Dionys and Daphne are involved in a mutual curing of illness, with Daphne visiting the Count, whom she initially tries to forget "as one tries to forget incurable things" (*Fox* 167), and doing little services for him as he convalesces. The Count, solicitous of Daphne's health almost from the first, returns the favour in a more Lawrentian fashion before he journeys back to Austria after the armistice. Her illness is meant to be a sign that she is paying too much attention to the outer rather than her inner world, but the pre-oedipal associations of illness for Lawrence, like those of food in *The Fox* and *The Captain's Doll*, get in the way of his attempts to revise the traditional hierarchies of his society in *The Ladybird*.[30]

Dionys's importance to Daphne is initially very similar to the significance the fox has for March at the beginning of *The Fox*, when the thought of the animal persistently "came over her like a spell" (*Fox* 12). After her first visit, Daphne would like to forget the disturbing presence of Dionys, but "[s]he never forgot him for long. He seemed to come into her mind suddenly, as if by sorcery" (*Fox* 169). Other occasionally used words like "spell" and "bewitched" add to the supernatural effect and help also with Lawrence's critique of Christian culture. This critique takes the form of belittling Lady Beveridge's traditional virtues, hope, faith, and charity (e.g. "'Then we can hope, dear'" [*Fox* 161]), but is realized more extensively through pagan imagery, as the name Dionys, of course, suggests. Dionys also, at one point, invokes a hammer of destruction that (along with the setting at Thoresway) suggests Thor's hammer in Nordic mythology. Perhaps Lawrence was thinking here of Madame Blavatsky's description in *The Secret Doctrine* (a work Lawrence knew) of Thor's hammer as "the magic weapon forged by the dwarfs against the Giants" (99) and thereby suggesting a David-and-Goliath-like inversion in the prevailing power structure (and giving added significance to Dionys's size). In the contemporaneous *Movements in European History,* Lawrence uses the image of the hammer to describe the force that destroyed the Roman order: "the Hunnish force was only the black hammer which smashed up the already broken Roman world, it did nothing toward building up a new world" (*MEH* 64). It should be noted that this was in the year 375, a few scant decades after the adoption of Christianity by Emperor Constantine. Generally, Dionys's pagan god of destruction is the principle opposed to Lady Beveridge's Christian God of kindness.[31]

In the halcyon days before the war, Count Dionys gave Daphne a thimble for her seventeenth birthday. It is an unusual object replete with Christian and pagan symbolism: "A snake was coiled round the base and at the top, for pressing the needle, was inlet a semi-translucent apple-green stone, perhaps jade, carved like a scarab, with little dots" (*Fox* 183). The snake and the apple are opposed to the scarab or "ladybird" of the title. The associations surrounding Dionys are sometimes ambiguous enough to be taken as either Christian or pagan, as in his temptation toward the end of the story to "take the soul" Daphne offers him into an underworld that is called both Hades and hell (*Fox* 216), but from the Christian perspective, the gift of the thimble with its apple-green stone and its snake suggests the temptation of Eve. Dionys, who claims at one point that "'I have a devil in my body, that will not die'" (*Fox* 169), now offers Daphne the secret knowledge of his society, and his presence certainly ends Basil's "heaven on earth" (*Fox* 183). Lawrence is retelling and inverting the Genesis story of the Fall with the Satanic figure offering a salvation of sorts rather than damnation. Lawrence, characteristically, is also condemning "modern" sex in the head, the self-conscious attitude that overtakes Adam and Eve in Eden. Dionys's armorial ladybird leads at one point in the narrative to a discussion of the Egyptian scarab (from which Dionys claims it descends) and thus to Fabre's idea that "'the beetle rolling a little ball of dung before him, in a dry old field, must have suggested to the Egyptians the

First Principle that set the globe rolling. And so the scarab became the symbol of the creative principle'" (*Fox* 209). In this way, Lawrence substitutes for the Christian Creation myth a pagan creative principle that incorporates its opposite, the concept of corruption and decay (the dung).

Like Hepburn in *The Captain's Doll*, Dionys has a wife, termed "unpleasant" (*Fox* 164), and two children, but Lawrence simply removes the marital relationship as an obstacle in one brief paragraph early in the story (*Fox* 168). Unlike both *The Fox* and *The Captain's Doll*, then, in *The Ladybird* there is no need to physically eliminate the blocking female character. Daphne's husband, Basil, is merely emasculated figuratively speaking, as the evolutionary conflict in this novella takes the form primarily of a sexual competition rather than one for raw survival. Here, Lawrence emphasizes the struggle for sexual selection, which Darwin describes in the *Origin* as depending "not on a struggle for existence in relation to other organic beings or to external conditions, but on a struggle between the individuals of one sex, generally the males, for the possession of the other sex. The result is not death to the unsuccessful competitor, but few or no offspring" (94). As noted above, there is even a brief reference in *The Ladybird* to Fabre (*Fox* 209), the same French naturalist Darwin cites with some evident amusement for his observation of "the males of certain hymenopterous insects ... fighting for a particular female who sits by, an apparently unconcerned beholder of the struggle, and then retires with the conqueror" (94). The description fits rather well the verbal war between Basil and Dionys at their first meeting, where Daphne has the satisfaction of knowing "that she was dangling the two men, and setting all the imprisoned officers wildly on the alert" and where she sits back silently observing the men "consumed together in a combustion of words," her sympathy with the Count and her intellect with her husband (*Fox* 197, 200).

After Basil begins to suspect that his wife has been smitten by the Count, "[h]e became suddenly ashamed to make love to her." He sees her now as somehow virginal, and, as for himself, "all the sex and the desire had gone. He didn't want it – he hadn't wanted it" (*Fox* 218). Through all this, Basil and Dionys remain friends, but only, presumably, because Basil has acknowledged his defeat by the foreigner in terms of sexual competition. Basil's only child with Daphne has been stillborn, and the novella implies that there will be no further offspring. The competition between the two men finally resolves itself in a manner similar to that in Lawrence's short story "The Border-Line," where Katherine's deceased first husband, Alan, reclaims her from the afterlife, although in *The Ladybird*, in keeping with the Persephone myth, Dionys is willing to share her. He tells Daphne: "'Now you are mine. In the dark you are mine. And when you die you are mine. But in the day you are not mine, because I have no power in the day'" (*Fox* 216).

One may also detect the Darwinian concepts of fitness and extinction in the more basic struggle for survival in *The Ladybird*. In the beginning, we read that in 1916–17 "the old spirit died for ever in England" (*Fox* 158), an extinction about which the narrator is somewhat ambivalent. Individually, Dionys is presented at first as someone who "might be dead," but he survives. As a wounded prisoner of war, Dionys has trouble accepting his continuing physical existence: "'Why have I survived into this?'" he complains (*Fox* 167). In a cancelled passage, he refers to the bitterness of the weak (*Fox* 260). As a matter of fact, something inside him has failed to survive, as Lawrence, in typical fashion, transforms what is conceived in Darwin as a struggle among individuals into a battle *within* the individual, where death and rebirth take on a peculiarly Lawrentian meaning. Nearing recovery, the Count informs Daphne that his happy self has been destroyed: "'That little animal is extinct'" (*Fox* 179). Since Lawrence's interest here is primarily on a sexual rather than a natural selection, however, fitness is modified into a term of the former struggle, as in Daphne's wish that the shirts she has sewn for Dionys had fit the flesh rather than the spirit (*Fox* 178). Dionys tries to convince Daphne with examples from the natural world that she is his fit mate. As for her legally- and socially-sanctioned mate, Basil has been pronounced "quite fit" after his own ordeal as a prisoner of war (*Fox* 184), but this statement proves ironic in Darwinian terms. As in the companion novellas, there is a determined effort on Lawrence's part in *The Ladybird* to place that conflict in the service of a redefined hierarchy that he offers as the solution to Europe's postwar malaise. Once again, however, pre-oedipal obsessions get in the way.

The first indication we have of a pre-oedipal concern that interferes with an effort to redefine hierarchy involves the view of Dionys as a dependent child. In Lady Daphne's initial visit to him, he is described as "small as a boy" (*Fox* 159), and Lady Beveridge later characterizes him as having "[s]omething of the terrible far-awayness of a child that is ill and can't tell you what hurts" (*Fox* 163). The Count is afraid that he is babbling like a child to the tall Lady Daphne: "'Ach! I have a head on my shoulders that is like a child's windmill, and I can't prevent its making foolish words'" (*Fox* 167). It troubles him that, as he says to his benefactress, "'I complain like a child, and ask for things. I feel I have lost my manhood for the time being'" (*Fox* 171). This is just after he has told Daphne that his hands hurt and asked her rather indelicately if he may use her hair as a bandage. In this way, the illness metaphor running through the novella is firmly connected to the question of sexuality, just as food and eating were so connected in *The Fox* and *The Captain's Doll*. Illness and dependency on female ministration will unman a man in Lawrence as surely as will hunger and the need for the nourishment only a mother can provide. The sexual competition we eventually see between Dionys and Daphne's husband, two wounded men in need of healing, in a way comes down to a question of which one of them can reverse the mother-child dependency on Daphne first or most effectively.

Dionys asks Daphne to replace his mother, as it were, in the business of the shirts. It irks this Bohemian aristocrat that he is reduced to wearing hospital-issue clothing. At home, his mother always sewed for him, "'[a]nd after her, my mother's sister who was the head of my house.'" It was never his wife (*Fox* 174). And now Dionys wishes Daphne to sew shirts for him. Even though the two eventually become lovers, then, Dionys places Daphne here in his mother's place, not his wife's. Why does Lawrence even bother to have Dionys tell us who it is that traditionally has sewn his shirts? It makes little or no difference to the storyline. In this detail, as in other otherwise extraneous or inappropriate scenes or details throughout the volume of novellas, Lawrence is patently writing under the compulsion of his unconscious needs. The personal needs at times control the writing and throw it askew from Lawrence's own intended purpose.

It is not simply that there is an occasional mention of or metaphoric use of mother and child figures in *The Ladybird*; the novella is replete with them. Lady Beveridge is called a "Mater Dolorosa" on the first page, an allusion that helps to get Lawrence from his compulsion to his intended critique of Christianity. The Ladybird itself is a "Marienkäfer," "[t]he beetle of Our Lady," Dionys explains (*Fox* 209). At one point, Dionys implies a likeness between Daphne and an ice-bear who hides with her cubs (*Fox* 179). When Basil returns to Daphne from his captivity as a prisoner of war, he finds her more mature than the "girl" he left behind. Now it is "'as if you were the moon-mother of the world'" (*Fox* 191). His awe and worship of Daphne is capped by his kneeling at her feet and sacramentally kissing them (*Fox* 193), much like Hepburn's reported kneeling in front of Evangeline on their wedding night in *The Captain's Doll*. For Lawrence, apparently, such a position is not only demeaning to the male but indicates a maternal dominance by the woman, even though one associates the action more with proposals of marriage than anything else in English culture. At the same time, Basil calls Daphne "a precious darling child" and a "perfect child" (*Fox* 194, 195) to confuse the issue, as Lawrence wants to suggest his culture is confused with regard to male-female relations. Basil's wish to die in sacrifice upon the altar of his worship of Daphne (*Fox* 195) is almost a parody of Dionys's wish to die earlier in the novella. Daphne herself feels like a "prostitute goddess" sullied by Basil's worship. Her feeling of sickness from Basil's love is a setback in her own recovery and a return in the text to the concern with illness. Indeed, at the very point of Basil's return, her near-nausea at looking at her "lilac-veined lids" over her eyes in the mirror (*Fox* 190) renders heavily ironic Basil's comment that Daphne's maturing into a woman has brought her out "like a flower" (*Fox* 191).

It is significant for the mother-child paradigm in the story that the singing that brings Daphne to Dionys's bedroom seems something of a reversion to childhood for the Count, who is about to be sent back to his home country. It is a "crooning to himself the old songs of his childhood ... in a small, high-pitched, squeezed voice" (*Fox* 212). Like the cooings of an infant, the words are incomprehensible to Daphne, but she finds them soothing. In a far too obvious and crude reversal of Basil's worship of Daphne, Lawrence has Daphne kneel at Dionys's feet in worship, and he now feels that he will be master of the life to come, "Father of the souls that would come after" (*Fox* 216), certainly not a child. The inversion is a culmination of what Lawrence has been carefully developing throughout the novella.

Dionys's argument for an inversion of hierarchy is oblique at first. He is now convinced that in a raving world, the only true sanity is madness (*Fox* 175). At a late meeting between Dionys and Daphne (just after he has been transferred to Voynich Hall from Hurst Place), Dionys confides in Daphne some of the mysteries of his "old secret society," a society that he compares to the freemasons and which sounds very much like a kind of theosophical association. Briefly stated, he is a member of an organization that believes that "'we've got the world inside out'" (*Fox* 180) and that to see the world properly would therefore, logically, require an inversion of what we believe we see. Dionys now subscribes to this doctrine. He sees himself to have been a strutting little society man, content to live from what "was just the inside-out of me" (*Fox* 181). Similarly, he argues that Daphne, whose sister-in-law bluntly told her that it was her business to be beautiful (*Fox* 184), must pay less attention to her outer appearance and, as it were, turn herself inside-out. He will find her root (*Fox* 182).

Apart from the obvious psychological compensation involved in a denial of apparent weakness, the advocacy of various forms of inversion is the scaffolding for Dionys's argument for a *social* inversion, even revolution, involving actual political power. It would be based on the notion of a natural aristocracy that Dionys sets out in some detail in his conversations with Basil. Kingsley Widmer, in viewing the way in which the argument for a new hierarchy is set forth in some of Lawrence's other writing but more specifically in *The Ladybird*, pinpoints what he considers to be "a serious contradiction in Lawrence's thought and a fundamental incoherence in his art. The dark hero is emphatically 'the outsider.' From where, then, his sense of power and authority? It comes … most importantly from the 'Kingdom of Death,' for he is 'King of Hades.'" Widmer goes on to ask a very good question: "Can the demonic outsider be a figure of social authority? Not, surely, if he is to remain the outsider, the rebel, the force of seduction and flame" (50).

While I agree that there is both contradiction and incoherence in the thought promulgated in *The Ladybird*, I disagree that its source is where Widmer places it. Widmer has erred in taking Dionys's notion of class, what the critic calls "the aristocratic principle" (49), in the conventional sense. If the evidence of this volume of novellas is to be taken into account as it must, Lawrence, at this stage in his career, wants to transcend conventional notions of class and to base merit on biological and spiritual superiority. This attempt may or may not be any more palatable than supporting an elite landed aristocracy, but the two are clearly not identical. That Lawrence in *The Ladybird* chooses as his major players figures from the ruling class in the traditional sense (the Lord and Lady Beveridge, Lady Daphne, Count Dionys) or from the political (if relatively impoverished) establishment (Basil Apsley)[32] makes Widmer's reading understandable. But from the evidence of the two companion novellas we get the strong impression that traditional notions of class are a false basis for the new Lawrentian hierarchy. Even in *The Ladybird* we clearly see the invidious effects of class consciousness most particularly in Daphne's fascination with the dependents at Thoresway and her inability to breach the chasm of class between herself and them: "Her *consciousness* seemed to make a great gulf between her and the lower classes, the unconscious classes. She accepted it as her doom. She could never meet in real contact anyone but a super-conscious, finished being like herself: or like her husband: or her brothers" (*Fox* 211).

Dionys argues that he and people like him should wield power "'[n]ot as a hereditary aristocrat, but as a *man* who is by nature an aristocrat … [I]t is my sacred duty to hold the destiny of other men in my hands and to shape the issue'"(*Fox* 202). Elsewhere, in the essay "Aristocracy," for example, Lawrence argues that the natural aristocrat is the one with more vitality than others: "One man *is*, in himself, more, more alive, more of a man, than another." Hereditary aristocracy is "bunk" and must give way to the biologically sanctioned natural aristocracy: "the whole of life is established on a natural aristocracy," which consists of those people who are able to provide "life," "release in us the fountains of our vitality" (*PII* 475, 477). This declaration accords with one of Lawrence's definitions of power in "Blessed Are the Powerful," where he writes that "power is the setting of life in motion" (*PII* 438). However, in *The Ladybird* – and it is here that I would place the ideological contradiction that Widmer observes – Dionys insists that he cannot fulfill his duty as a natural aristocrat "'till men willingly put their lives in my hands,'" and that will only occur when "'men who are really living [come] beseeching to put their lives into the hands of the greater men among them'" (*Fox* 202), a notion very similar to what Lawrence himself states in a letter to Cecil Gray (*L* iii 262). So, according to the "logic" of the argument, those choosing the leader will be high on the scale of vitality to begin with, the "really living." Their goal is to find someone who can create vitality. But since they cannot choose him (or her?) unless they are vital to begin with why do they need a leader to create what they already have, and where does this leave the non-vital people? There is, of course, a subtext here directed toward Daphne. On the interpersonal level, Dionys is suggesting that, at a certain point, Daphne will put her life in his hands and that it will be because of the vitality within her, and she does. And when she does, as we have seen, it is a culmination in a subtextual effort to invert the dependency of child upon mother, the pre-oedipal bond.

A second and related biographical factor is relevant here. Lawrence believed that those who rose to prominence during the war, like Lloyd George, were despicable from many points of view. In *The Ladybird*, this feeling is expressed through the person of Lord Beveridge, who wishes to "defy scandal" in allowing the enemy officer to stay at his estate for a fortnight: "It was almost the worst nightmare of all, this coming to the top of a lot of public filth which was determined to suffocate the souls of all dignified men" (*Fox* 205). This "filth," this "unclean scum" are also termed "mongrels." England has "fallen under the paws of smelly mongrels" (*Fox* 205, 206). If the "natural aristocracy" upon which the whole of life is established is to claim its rightful place, it will have to force back the mongrel herd from the top of the social and political hierarchy. Small wonder, then, that in *The Ladybird* Lawrence provides a biocentric argument as still another subtext in the quest for hierarchical inversion.

Dionys has some marked biocentric opinions. He cannot abide flowers that are pillaged from their state of nature in order to create "upholstery for human life" (*Fox* 168). He rejoices that the world "'is not only a world of men'" (*Fox* 172), the very world he claims later he would have destroyed. Since Daphne does not understand his biocentric point of view, she cannot understand what drives him. Nevertheless, he is the major advocate of biocentrism in the novella. He is often likened to animals of various sorts and, despite his claim to Lord Beveridge that, as a prisoner, he has discovered that he is "*not* truly a beast" (*Fox* 207), he often compares himself to animals, as does Daphne in her own mind. As a squirrel, Dionys is able to take his nourishment from the natural world, thereby lessening his dependence upon woman. The chestnut he eats is "[l]ike sweet milk" (*Fox* 185), the maternal nourishment *par excellence*. Daphne's "mastering" of Dionys at dinner at Thoresway turns him into a "sacrificed beast" whose entrails are read as auguring the future (*Fox* 208). His "call" to Daphne is "utterly inhuman" (*Fox* 214). For her part, Daphne tries for a time to place Basil in the biocentric word: "Like an ear of corn he was there for her gathering – her husband, her own lovely English husband" (*Fox* 182). But all Basil can be, as he himself suggests in a letter to Daphne before his release, is "a starved rabbit" and "a miserable brute" (*Fox* 183). The basic, biocentric premise is that if a just and rightful natural aristocracy cannot be created then humanity as a species is doomed, and that is all right because something better will come after. If there is "the quick of a new thing" that Lawrence claimed for this novella, it involves this biocentric point of view. However, as the evolutionary argument is not much different from the one at the close of *Women in Love*, the "new thing" (not to mention the "quick") may be something of an exaggeration.

In the end, the journey toward leadership as the means by which to ensure a transformation of society is blocked by the obstacle of a compensatory individual transformation. Dionys feels that things are happening in his own soul (*Fox* 199); Daphne becomes "different" after Dionys's love. The ending rings false perhaps because we sense that even in personal and interpersonal terms things have advanced only from the pre-oedipal concerns of dependency to an oedipal triangle that is but a poor retelling of the Pluto-Persephone myth. Dionys's final advice to Daphne is: "'If you have to give yourself to your husband, do so, and obey him. If you are true to me, innerly, innerly true, he will not hurt us'" (*Fox* 220).

Sandra M. Gilbert has argued that we see in Lawrence "a defensive movement toward devaluation of the female" along with "an acknowledgment of the male subject's dependence on female, and specifically maternal, power" (139). She views *The Ladybird* as a good example of Lawrence's ambivalence toward female power, and, as part of the evidence for her view, she adduces the story of the swan woman and swan king with which I began this discussion. The use of the swan story leads Gilbert to suggest that, in addition to the Greek myths Lawrence was obviously drawing on in his novella, he may also have had in mind a Grimm fairy tale by the name of "The Six Swans," for Gilbert, "a story about female power" (148). There are, in fact, a number of parallels between *The Ladybird* and "The Six Swans," particularly the motif of sewing shirts to effect a cure.

It has been speculated that the fairy tale is traceable to the custom of ultimogeniture common in eighteenth-century Hessia, a non-patrimonial inheritance system through the youngest daughter, who held property in trust for the family, but a system that in reality left the father in control (Zipes 48).[33] In the Grimm tale, there is a witch who, like a kind of evil Ariadne, leads a king who has lost his way out of the forest in return for his taking her daughter as his queen. The king has six sons and a daughter from a previous marriage, and, not trusting his new wife, he hides them in a forest abode to which he can find his way only with the help of a magic ball of yarn given to him by a second Ariadne figure, a wise woman. Predictably, the wicked queen discovers the children and turns all of the boys into swans by sewing magic shirts that she throws over them. The daughter flees and discovers that in order to restore her brothers she must sew them six shirts of asters over the course of six years, during which time she is not to utter a word. She is discovered in the forest at her task by another king, who weds her because of her great beauty. However, at each successive birth of a son, the king's own evil mother (a doubling of the forest witch) takes the child from the queen, smears her mouth with blood and accuses her of killing her own children and eating them. The queen will, of course, say nothing in her own defence until the six years are over. That occurs on the day she is to be executed for murdering her own children. On that day, her brothers come flying to her as swans, she restores them with the shirts she has sewn, and she denounces her wicked mother-in-law.

Gilbert underplays the negative associations with female, especially mother, figures in the Grimm tale. The story ends with the king ordering the gruesome execution of his own mother, who "was tied to the stake and burned to ashes" (Grimm 185). We are in the territory of "gynocide" here, as we were in *The Fox* and *The Captain's Doll*. This possible intertext of *The Ladybird* serves to underscore the apparent fact that the path to the revitalization of society by means of male leadership is littered with the dead bodies of women. Certainly, as Gilbert says, Lawrence was "[h]aunted by female primacy" (150), but she gives him more credit than he deserves for finding a resolution that involves an ultimate acquiescence in the power of the mother (161). The fictional project of his 1919–22 period, his almost desperate use of Darwinian concepts in constructing a theory of vitality completely removed from generation, is a testament to that fact.

The *Ladybird* novellas bring Lawrence to a temporary dead end in his efforts to redefine the "love ethic" (Spilka's term) of his marriage works in terms of his developing ethos of leadership. The attempt in the volume to rethink hierarchy has much in common with the carnivalesque as defined by Bakhtin in *Rabelais and His World*. Carnivals for Bakhtin celebrate the "temporary liberation from the prevailing truth and from the established order; they marked the suspension of all hierarchical rank, norms and prohibitions" (10). And if Bakhtin is right that what he calls the grotesque body "is always transgressive, disclosing its essence as a principle of growth which exceeds its own limits only in copulation, pregnancy, childbirth, the throes of death, eating, drinking, or defecation" (26), then Lawrence's bid to work out the challenges he has posed himself in the *Ladybird* novellas might well be expected to take a carnivalesque and grotesque turn. This, more than anything else, may explain the introduction of the Natcha-Kee-Tarawa carnival troupe into the next major work of the transitional period, *The Lost Girl*, which we will examine in the next chapter together with the period's other completed novel, *Aaron's Rod*.

3 Confinement and Survival in *The Lost Girl* and *Aaron's Rod*

In the transitional fiction we are concerned with, there is generally a bi-level narrative structure. On the "upper" level, Lawrence's intentions are worked out consciously through some variant of his own version of the Sleeping Beauty tale in conjunction with a number of evolutionary concepts. On a "lower" level, like a hidden current flowing against the waves, the author's intentions are subtly subverted by survival anxiety that is largely unconscious and that we are able to perceive only by cracking the code signalled by inappropriate imagery or plot turn. L.D. Clark argues that the novels between *Women in Love* and *Lady Chatterley's Lover*, works he refers to as Lawrence's "pilgrimage" novels, partake of the romance characteristics of the early American novels Lawrence was reading and writing about at the time.[1] In fact, the very doubleness in Hawthorne, Melville, Poe, and the others that Lawrence proposed was generated by a conscious scheme working against an unconscious imperative – so much so that we must, to paraphrase Lawrence's justly famous advice, trust the tale rather than the artist in our reading – is present in his own writing of this time. H.M. Daleski's remark in a different context is appropriate to the entire period: "Lawrence … in terms of his own division of the world, is divided against himself" (*Forked Flame* 210).[2] One could even maintain that the bi-level structure of these works is one source of the critical disagreement about the meaning, even worth, of the fiction. So, for example, Gary A. Wiener contests Hilary Simpson's view, in her *D.H. Lawrence and Feminism*, of the Alvina-Ciccio relationship in *The Lost Girl* as oppressive (Wiener 248), but much of his evidence is doubtful in that the statements he quotes from the Hardy study are those Lawrence would no longer make in the face of the severe challenge to survival represented by the unconscious anxieties at work in *The Lost Girl* and other works of the transitional period. Wiener is responding largely to Lawrence's manifestly plotted design in the novel, while Simpson senses an underlying hostility to the female protagonist whose life the text is ostensibly championing.

To summarize from the last chapter, in the three novellas, Lawrence's use of food and illness motifs triggers survival concerns at the subversive level of narration that flows under the stories of physically repressed or developmentally rigidified women (or man in the case of *The Captain's Doll*) who are sleepwalking through life until a charismatic "prince" awakens the Sleeping Beauty to a life of awareness through the senses. That the projected fulfillment in these stories is heavily qualified at the end speaks to the power of the subversive narrative, which, by the conclusion, has alerted Lawrence to the fact that something is not quite right in his intended design. Thus we have the noncommittal endings as the author awakens from his own sleepwalking through an ideology-driven narrative that simply will not do because of his deep anxieties about survival. In "The Spirit of Place," Lawrence's terminology describing the doubleness of American literature resonates with implications for his own metafictionality: "The author is unconscious of it [the duplicity in American literature] himself. He is sincere in his own intention. And yet, all the time, the artist, who writes as a somnambulist, in the spell of pure truth as in a dream, is contravened and contradicted by the wakeful man and moralist who sits at the desk"(*SM* 18).[3]

The survival tropes of food and wellness/illness we saw in the novellas are also present in *The Lost Girl* and *Aaron's Rod*, the two completed full-length novels of the transitional period (*Mr Noon* having never been finished). In *The Lost Girl*, for instance, one of Woodhouse's established commercial activities is described as "the lustre of lard and butter and meat," the apex of society in this Midlands town is "the rich and sticky cherry of the local coal-owner," and one persistent social problem the "Dead Sea fruit of odd women, unmarried, unmarriageable women, called old maids" (*LG* 1). And on it goes. But in both novels a larger system of imagery, one that has been present in the novellas but less developed, takes precedence and clearly points to the underlying survival anxiety. This is the imagery of confinement. In the essay "On Coming Home," Lawrence expresses his own horror at the feeling of confinement on returning to England after years abroad: "That is how one feels. Like a tiny porcelain figure shut in inside box after box of repeated and intensified shut-in-edness. It is enough to send one mad" (*PII* 253). In *The Fox*, Henry feels the fox's doom, a wild thing (like himself) netted in by the "civilized" life of England. In *The Ladybird*, Dionys is a prisoner of war confined to a military hospital. In *The Captain's Doll*, Hepburn looks out to the moon to escape a constricting marriage whose qualities he projects onto the image of the bird cage. In *The Lost Girl* and *Aaron's Rod*, we see a pattern where some form of confinement prompts a response that, in turn, leads to pre-oedipal patterns of anxiety regarding survival that undermine the ostensible drift of the narrative. In *Aaron's Rod*, as we shall see, the pre-oedipal pattern is manifested particularly in imagery of breathing, a function that is, obviously, basic to survival; in an odd cannibalism trope; and in a tendency to infantilism, or a drive to return to the mother that undermines the surface theme of male self-sufficiency. In *The Lost Girl*, confinement and illness form the dominant subversive pattern that threatens Lawrence's scheme of inversions by which he attempts to suggest his protagonist's salvation. We begin with this latter work.

THE LOST GIRL: FROM COFFIN TO CAVE

Lawrence may have chosen a female protagonist in *The Lost Girl* because he wished to "rewrite" Arnold Bennett's *Anna of the Five Towns* (1902), as a number of commentators have suggested,[4] but the narrative strategy of using a female perceiving consciousness whose own survival depends upon the sexual selection of a mate clearly also serves the compelling psychological needs of the author. The loss of self that is part of the significance of the novel's title can be contemplated in a positive way at this point in Lawrence's career only with respect to the female of the species. Although there is always something of Lawrence himself in his protagonists, male or female,[5] any identification between Alvina and the author serves the ideological purposes of the period. In other words, Lawrence uses Alvina as a symbolic means of purging from himself his own unwanted weaker traits in the project of survival. Narrative complications arise, however, when the ruthless elimination of blocking characters – in this case, the heroine's mother, mother substitute, and father – seems about to extend to the heroine herself in a more than metaphoric sense by the end of the novel. In terms of narrative strategy, the implicit and impossible-to-answer question that remains at the end is this: if Alvina has done all the right Lawrentian things to become the rescued Sleeping Beauty, has made all the right sexual choices, why then is she threatened with renewed sleep and even extinction at the conclusion?

Alvina Houghton is the only child of James and Clariss Houghton in the Midlands town of Woodhouse, whose northern location (relative to Italy even more so than to London) suggests the coldness in human relations that the protagonist, unconsciously at first but then with a wavering conviction, longs to replace with the warmth of human love somewhat in a Forsterian way (*Where Angels Fear to Tread* is sometimes mentioned by critics as an influence). Her governess is named Miss Frost, so the intention is crystal clear, so to speak. Alvina's first serious attempt to find warmth involves a small, dark man, Alexander Graham, a newly graduated doctor from Australia, who is described by the narrator in decidedly physical terms. Unfortunately for the heroine, he eventually loses the battle with Miss Frost for Alvina's heart. With spinsterhood threatening, Alvina decides to train as a maternity nurse, another indication of her half-conscious desire to align herself with life forces. In the meantime, James Houghton, originally a merchant selling "Manchester goods" and other fabrics, is engaged in increasingly dubious business endeavours including the manufacture of ready-made clothing for ladies, a development that brings a Miss Pinnegar, James's manageress for the work-girls, into the family orbit. A later project is the attempt to extract coal from a disused mine, and Alvina's second major encounter with dark physicality occurs when she visits the colliery works and "is wound down in the iron bucket to the little workings underneath" (*LG* 46) to visit the colliers toiling underground. She feels that "something bubbled up in them, the force of darkness which had no master and no control" (*LG* 48). Another suitor, Albert Witham, comes and goes for Alvina, and then there enters into Woodhouse a carnivalesque force of vitalism that will forever change Alvina's course of life, the Natcha-Kee-Tawaras, a Fenimore Cooper-inspired theatrical troupe of mixed European origins who purport to bring to provincial England something of the flavour of the American Wild West. The carnivalesque formlessness of the troupe may be a self-reflexive signature suggesting the fictional form that contains it.[6] James, with the help of the droll theatre man Mr May,[7] comes into contact with them through his last endeavour, a cinema house, which must have live theatrical "turns" to fill out the film program. Among the troupe is an Italian, Francesco "Ciccio" Marasca, who, although he is part of an artificially primitive show, will bring Alvina the instinctive flow of life she has been looking for and eventually take her off to a savage part of Italy. Before that, however, Alvina has one more stab at English respectability. When her father dies, she leaves Woodhouse for a job as maternity nurse in Lancaster. Here she gets involved with Dr Mitchell, essentially an Albert Witham with a stethoscope. Alvina gets engaged to the good doctor, and only a chance meeting with Ciccio alters her determination for bourgeois conformity. She marries the Italian, five years her junior, and submits "mindlessly" to his narcotic spell. Alvina's last view of England from the boat that will take her and Ciccio to Italy shows "corpse-grey cliffs" and an "ash-grey coffin" (*LG* 294), a description that reinforces the text's symbolic import and suggests an unflattering significance to the name "Woodhouse" and what it connotes of death and confinement. The novel ends in a suspended narrative movement, as the war intrudes and Ciccio must go off as a soldier while a pregnant Alvina awaits his return.

Given its on-again, off-again history of composition (a seven-year hiatus intruded between drafts), it is not surprising that *The Lost Girl* is an incompletely realized work of fiction, although it was widely praised by contemporaries, including Bennett, for its provincial realism.[8] Lawrence's intentions certainly seem clear enough. The novel is about a non-religious, even Blakean, form of salvation and the threat to salvation posed by various forms of confinement and illness that reflect the dangers of English culture.[9] The familiar Lawrentian plot is – supposedly – reinforced by metaphorical devices. Through the characteristic device of inversion, he wishes to show the deadliness of English culture and the life-enhancing qualities of a more primitive way of life. He will do this through the life history of the "lost-girl" protagonist of the novel, who is able to find love and fulfillment, like *The Rainbow*'s Tom Brangwen, only with a foreigner (the radical Other).

The overt narrative strategy of the novel is evident in this carefully developed pattern of inversions. There is, most obviously, the inversion of the usual connotations of ascent and descent so that, in Julian Moynahan's words, Alvina "extricates herself from drab circumstances not by rising above, but by sinking below them" (122), the Lawrentian version of the Fortunate Fall whereby a "fallen" woman is a fulfilled one. And, as the title intimates, we have the paradoxical notion that to find oneself one must become lost. Worthen suggests that Lawrence's use of "lost" in this novel may be a deliberate, anti-Victorian rejoinder to Dickens' portrayal of the morally lost Emily in *David Copperfield* ("Recovering" 214–17). We see Alvina's confinement in the coffin of Woodhouse (housed in the wood of a coffin) inverted in a preliminary way in her work as maternity nurse, where she presides at the "confinement" of others. We witness the deathliness and ashes of James Houghton's mining endeavour, the "Throttle Ha'penny" (a near homonym to "throttle happiness"), reversed in the vitality Alvina finds underground among the working men. She must descend in order to develop onwards, a paradox that clears the way for her eventual descent in class terms and in her relationship with Ciccio, who, like so many of Lawrence's males in this period, has a grudge against mothers.

There are also inversions in terms of the child motif, always dangerous territory for Lawrence. Dr Mitchell, perhaps Alvina's most ardent English suitor, treats his patients like children. But he becomes infantilized himself in his relationship with Alvina, so much so that she must virtually flee from his dependency. By way of contrast, in Alvina's relationship with Ciccio, he appears childlike at first, his touch "almost like a child's touch" (*LG* 127), and Alvina can all but hear Mr May say of Ciccio, "'Like a child, you know, just as charming and just as tiresome and just as stupid'" (*LG* 138). But it is Alvina who, in the end, becomes a child in dependency. After Ciccio has "killed" her with his sexual insistence, "she pressed her face on his chest and cried, he held her gently as if she was a child" (*LG* 202). Characteristically, Dr Mitchell hesitates at the thought of having children with Alvina after their planned marriage, since a child would displace him in terms of infantile dependence on the mother: "A child would take her away from him" (*LG* 262). Ciccio, on the other hand, is quite pleased at the prospect of being a father (*LG* 321). Part of Dr Mitchell's attraction for Alvina seems to be as a father-figure.[10] Alvina has just lost her woefully inadequate father to death; he has failed in his last will and testament, as in everything else, to take care of her. She meets Dr Mitchell, who appears to offer the kind of care and fatherliness she has always lacked, but when he turns out instead to be a big baby, she is horrified.[11]

As is usual in Lawrence, light and dark imagery creates a significant set of inversions. Alvina has been wrapped up in the forms of deathliness all of her life, and, as a consequence, has difficulty in recognizing life when she sees it. Alexander Graham is, like Count Dionys Psanek of *The Ladybird*, a dark, little man who, Miss Frost claims, has "dark blood in his veins" (*LG* 23). Alvina is too unsure of her direction in life at this point to withstand the distaste of her father and Miss Frost for this dynamic foreign man. Her indecision, very similar to that of March in *The Fox* regarding Henry and that of Hannele in *The Captain's Doll* regarding another Alexander, is signified through tropes of light and dark and the dichotomies of small and large, sanity and madness. The issue for Miss Frost is "clear as daylight," as it is to Alvina only in "her periods of lucidity." At these times, Alexander seems little, an outsider and inferior. However, at other times, like a Persephone flitting between a light and dark existence,[12] her daylight mood gives way to a nighttime vision that seems suspiciously like the compensatory fantasy of an impotent male, where "the little man loomed large, terribly large, potent and magical" and Alvina "felt she was going out of her mind" (*LG* 24). For Lawrence's protagonist, going out of her mind at this point is almost as good as getting out of Woodhouse. But because Alexander has gone on to Australia to wait for her, Alvina, again like the heroines of the novellas, when lacking the physical presence of their men, decides against him. Little does she realize that her survival is put in jeopardy by her decision. As Ciccio later says, suggesting more than he intends, "'You can live without food, but you can't live without fire'" (*LG* 220). Survival in something more than a zombie state, suggests this text, depends precisely upon making the right choices in life, for fate can be predatory: "'Answer or die,' said the Sphinx of fate ... Alvina's sphinx was an old, deep thoroughbred ... [a]nd her thoroughbred teeth were long and sharp" (*LG* 243).

The most developed form of inversion in *The Lost Girl* is that of ascent and descent, which is often coupled with light and darkness. Alvina has already missed her opportunity with the dark Alexander Graham to realize that the way up is the way down under. Subsequently, she embraces the liberating darkness of the mine shaft as she accepts the positive aspects of the "dark" reality of physical life that terrify her frigid and astringent female models, her mother and Misses Frost and Pinnegar (cold and vinegar). Although she feels entombed and frightened underground, Alvina believes that there resides the power of some "ponderous and inevitable" secret knowledge, in effect, Lawrence's ponderous and inevitable "blood" knowledge, against which "her mind dissolved" (*LG* 47). Thereafter, like Dionys of *The Ladybird*, Alvina is convinced that what the world needs is not "a new Jesus, another Saviour from the sky ... [but] a Dark Master from the underworld" (*LG* 48). That the dark and physically masterful Ciccio Marasca turns out to live high up in the Alban Mountains is one of the ironic self-subverting elements in the novel.

When Ciccio pays a call to the Houghtons to inquire, on behalf of the Natcha troupe, after the health of James (who has just died as it turns out), Alvina "stood on the doorstep above him" and looks down on him "from her height." Nevertheless, Ciccio beckons with his eyes, in which there is "a dark flicker of ascendancy," and Alvina steps "down to his level" to follow him (*LG* 174). When he kisses her, Alvina paradoxically comes alive even as she feels herself die, and, to complete the ascent-descent inversion, when Alvina runs indoors to her room, in her realization that she loves Ciccio, she "kneeled down on the floor, bowing down her head to her knees in a paroxysm on the floor" (*LG* 175). Alvina is beginning to embrace the inversion of societal norms. As she says to the leader of the Natchas, Madame Rochard, later, "'Why should things always go up? Why should life always go up?'" (*LG* 183). She has relapses along the way, however, and breaks with the Natcha group shortly after one bout of regression to old values that is signalled by the reversion to conventional images of ascent and descent. Her sense of pride "mounted obstinately" as she feels that "she was demeaning herself shamefully" in consorting with the group. She "had been bred up different from that. They had horribly low standards … Really, she had come down in the world" (*LG* 209).

With Dr Mitchell, the relative positions are very different, and that may be part of the attraction for Alvina since she has suffered through "The Decline of Manchester House" (the first chapter) in her childhood and "The Fall of Manchester House" (the tenth chapter) after her father's death and the discovery of his debts. After a time, the consciously elevated Dr Mitchell deigns to step down towards Nurse Houghton: "From his imperious condescension he took to a tone of uneasy equality … and little by little he sank" (*LG* 257). One of the benefits, in conventional social terms, of Alvina's engagement to Dr Mitchell is that "it elevated her in the common esteem tremendously" back home in Woodhouse. She is moving up in the world once more, although, in the northern town of Lancaster where she works, she feels the engagement somewhat diminishes her individuality (*LG* 273). In the end, the "Honorable Engagement" (as the relevant chapter is called) will not work because what Alvina wants most has nothing to do with honour or social standing. All it takes is the encounter with Ciccio to remind Alvina of her priorities and to reduce Dr Mitchell to a grovelling level of dependency that belies his social ascendency. When Alvina clearly indicates it is over between them, this imperiously paternalistic man acts "like a small boy," and "he fell on his knees before her" (*LG* 268) in what functions as a parody of Alvina's earlier prostration in her realization of her love for Ciccio.

The ultimate journey with Ciccio to his southern Italian home from this northern English town is fraught with paradoxical imagery of ascent and descent. In what is clearly a descent for Alvina in terms of social standing and class and even in terms of level of civilization in this "wild" and "savage" place, they travel ever upward into the hills that become a symbolic landscape as the heroine travels "up the darkness" (*LG* 304). Gradually they leave behind the trappings of civilization: the boat, the train, and the bus give way to a cart, an ass, and finally no road at all. Only in descending in several ways, in losing herself, will Alvina be able to ascend to fulfillment, to find herself, to awaken. But in this awakening, Lawrence makes clear, there is an irony, since the female Sleeping Beauty in all of these stories must ultimately be put back to sleep, as if the Grimm tale were reversing itself after Prince Charming's kiss has done its work.

The theme of awakening was important to Lawrence's conception of his novel early on, as the surviving fragment from the early, first-person version of *The Lost Girl* known as "Elsa Culverwell" indicates. Elsa thinks of her mother as "waiting to be awakened, fearing almost to sleep all her life, only tortured by a dream of unfulfilment" (*LG* 343). In the published version, Alvina goes through a series of awakenings, and all of them are viewed positively in the narrative until the last one. For instance, in her relationship with Dr Mitchell, Alvina suddenly realizes the absurdity of her intention to marry such a man simply in order to be able to continue to sleepwalk comfortably through life: "Her spirit, which had been as if asleep these months, now woke rather irritably, and chafed against its collar. Who was this elderly man, that she should marry him?" (*LG* 265). However, at the primitive home of Pancrazio, Ciccio's uncle, Alvina has an awakening that now must be resisted. She "woke with a start" and "[a] real terror took hold of her" to the point where "[h]er very being was threatened" until she crosses over to Ciccio in his straw bed and "felt his power and his warmth invade her and extinguish her" (*LG* 313). The long road to self-fulfillment seems about to end in self-extinction, and this for a protagonist that the narrator claims is an extraordinary individual (*LG* 83). The only trouble is that this individual happens to be female.

That gender is the issue here is made clear by the portrayal of the hyper-conscious Mrs Tuke, at whose confinement Alvina presides as maternity nurse and at whose home Ciccio serenades Alvina. To Mrs Tuke, life is a Darwinian nightmare and childbirth a threat to her human self-awareness in that it turns her into a howling animal. The universe is a machine and "life is a mass of unintelligent forces to which human beings are submitted" (*LG* 283).[13] While Ciccio's face in the same chapter is described as "beautiful, non-human" (*LG* 278), there is a crucial distinction for Lawrence between non-human but living on the one hand and mechanical and non-living on the other. Mrs Tuke will not submit her human, female individuality to the greater forces of life. Alvina must awake from dead forms, but she is not to place her female individuality above all else. Mrs Tuke seems to represent something similar to what Hermione and the Bohemians suggest in *Women in Love*, that is, a modernist and decadent overemphasis on the mind and the will at the expense of the instincts.

Alvina's personal struggle is defined by forces at work in society at large, and the society from which she (like her creator) has fled is slowly drifting toward its demise. It is not fit to survive. There is one vivid scene that, in effect, allegorizes the rot of England and the danger it poses. At this point, Alvina is attracted to a Woodhouse tradesman by the name of Arthur Witham (whose educated brother, Albert, becomes one of Alvina's rejected suitors). Arthur is repairing the organ in the chapel while Alvina is arranging the flowers when the rotten steps Arthur has been standing on give way and nearly break his leg. Alvina goes to his aid, and Arthur presses her hand down into his bloody wound for a long moment so that her hand becomes soaked with his blood. Rot leads to a collapse that only "blood knowledge" can ameliorate.[14] The telltale symptoms of incipient collapse that are emphasized throughout *The Lost Girl* are gender confusion, illness, and the cinema.

The gender confusion begins with James Houghton, whose actions and character are clearly meant to be pregnant with meaning. He is the most fully developed character in the novel after Alvina. One of his early "endeavours," after years of decline in his fabric business, is ladies' ready-made clothing (potentially an unintended irony suggesting Lawrence's tendency at this point to force his female characters into ready-made, ideologically tailored roles). The girls who work for James sewing the clothing "proclaimed the fact that he tried-on [*sic*] his own inventions upon his own elegant thin person before the privacy of his own cheval mirror. And even if he did, why not?" (*LG* 10). Throughout the novel, James Houghton is associated with machinery or mechanisms of one kind or another. The implication seems to be that industrialism (the machinery) leads to social decadence (the cross-dressing).

In James's "last endeavour," the cinema house, films alternate with dramatic "turns" booked by the music-hall agent Mr May and performed by travelling vaudevillians for the most part, among whom the Natcha-Kee-Tawaras are the most significant for the narrative. One of the turns is "a comic drama, acted by two Baxter Bros. disguised as women, and Miss Poppy disguised as a man" as the Bakhtinian concept of the transgressive body comes to expression in this carnivalesque scene that is one of several (*LG* 110). When Madame is too ill to perform, Mr May offers, perhaps a little too enthusiastically, to play her part in the theatre: "'I'll be the squaw'" (*LG* 133). And he does, although he turns out to be "a little too frisky as Kishwégin" (*LG* 136). Miss Pinnegar thinks Mr May "'exactly like a girl. Well, well – it makes you wonder what is and what isn't'" (*LG* 141). We must not mistake Lawrence's intentions here: he is not interrogating the construction of gender differences in a chauvinistic society on the edge of great changes in gender roles brought about by the war. He is, rather, suggesting the decadence of his own society in its failure to better distinguish between men and women and their proper roles. Ciccio's most basic function is to restore gender order at least in Alvina's life if not in Woodhouse. But this is dangerous territory for a man like Lawrence who is in doubt about his own gender identification. Indeed, in a metafictional sense, one might say that the cross-dressing begins with the choice of a female perceiving consciousness by a male writer.

Similarly, the pervasive illness motif, as in *The Ladybird*, is ostensibly a mark of a sick society with a kind of death drift, but underneath the satire of characters like the hypochondriac Mrs Houghton lies a real anxiety regarding Lawrence's own robustness and ability to survive. The victim of a loveless marriage, Mrs Houghton makes a domestic career out of invalidism somewhat like Lady Bertram of Austen's *Mansfield Park*, although she is less indulged.[15] Her illness – heart trouble, inevitably – is charted until her barely mourned death very early on in the book. The narrator blames her heart disease and early death on modern women's expectations of happiness, a central issue as well in *The Fox*: "Why should anybody expect to be *made happy*, and develop heart-disease if she isn't? Surely Clariss' heart-disease was a more emphatic sign of obstinate self-importance than ever James' shop-windows were" (*LG* 43–4). The search for happiness or the quest to make others happy, indeed, infantilizes its adherents in Lawrence's transitional fiction. The narrator likens "a mankind howling because it isn't happy" to "a baby in the bath" (*LG* 44). The self-parodic rainbow with the pot of happiness at the end of it that occurs in *The Fox* reappears here somewhat bawdily in one of Miss Poppy's dances: "Up goes the curtain – Miss Poppy twirling till her skirts lift as in a breeze, rise up and become a rainbow above her now darkened legs. The footlights are all but extinguished" (*LG* 110).

There may be something more disturbing going on in the description of Clariss Houghton as well. Lawrence, like Woolf in *Mrs Dalloway* (1925), alludes to Samuel Richardson's violated tragic heroine in choosing the name of his character. In fact, we know that Lawrence wished to consult Richardson's *Clarissa* (1747–48) at about the time he was finishing as much of *The Lost Girl* as he would manage before the war (Herring 18). As in *Mrs Dalloway*, the twentieth-century Clarissa figure is a victim of a loveless marriage in something of an inversion of Richardson's eighteenth-century tome. In Woolf's novel (where Clarissa also has a weakened heart), the frigid state of her marriage seems to suit Clarissa Dalloway just fine for the most part. In Lawrence's work, it is virtually the cause of her death. This woman's survival depends upon a man's love that is not forthcoming. Mrs Houghton keeps to her bedroom all the years of her marriage after Alvina is born, a Sleeping Beauty who never awakens because her Prince Charming is busy with money-making schemes as compensation, in effect, for the money she did not bring to the marriage. (James was expecting his wife, the daughter of a squire, to bring ten thousand pounds with her to the marriage, but she came with only eight hundred [*LG* 2].) That she is sexually frustrated is indicated by her reaction to the information that, in James's Sunday school class, order is maintained by a "tall and handsome" blacksmith (a variant on the Lawrentian gamekeeper or gypsy figure) who squeezes the knees of recalcitrant boys in order to strengthen the hand, as it were, of the ineffectual James. This "secret so haunted the poor lady that it marked a stage in the increase of her malady, and made Sunday afternoon a nightmare to her" (*LG* 9). For his part, James channels all of his sexual energy into his schemes, one of which is for a private hotel that is to be "an erection on a higher plane" (*LG* 56). Conversely, even Ciccio's unwanted love is presented as salvational to Alvina. In what is a virtual rape scene, the daughter of Clariss becomes a Richardsonian Clarissa but in a work where a fallen woman is a saved one, as the descent imagery culminates.[16] Ciccio simply seizes Alvina and makes love to her against her will, after which she finds his beauty to be "forever unseizable" and must fight a desire "to fall at his feet" (*LG* 233–4). As Phillip Herring suggests, "Ciccio responds to her love by attacking her self-respect as a prelude to the surrender his culture and male pride have led him to expect of a wife" (13). It is more than Alvina's self-respect that is at stake here, however; it is her very selfhood.

Lawrence also uses Clariss Houghton to undermine the power that mothers hold in his own mind. Such a weak and ineffectual woman can hardly be a threat, and Alvina barely thinks of her after her death: "So Alvina refrained from pondering on her mother's life and fate. Whatever the fate of the mother, the fate of the daughter will be otherwise" (*LG* 43). And that of the son as well, we are tempted to add. Mrs Houghton is an interesting example of the contradictory way Lawrence will virtually at one and the same time deny and admit the power of women/mothers. This mother has no influence on her child; at the same time, Alvina's "real" mother, Miss Frost, has far too much and so must be killed off, punished even, just as Banford and Mrs Hepburn had been (and as Lady Beveridge had been socially marginalized in a more metaphorical kind of death).[17] And what an excruciating punishment it is, for Miss Frost must be made to thaw out just enough to feel the same pain of unconsummated love that she has foisted upon her "daughter." Very soon after the death of Mrs Houghton (seven pages later to be exact), Miss Frost simply fades away like a snowflake in July. She develops pneumonia after a man she has been seeing is transferred by his firm to another location (*LG* 49–50). As with the death of Banford in *The Fox* and Mrs Hepburn's fall in *The Captain's Doll*, the narrative strategy here is at the service of the protagonist's vitalistic survival instinct. For, much earlier than the death itself, the narrator informs us that Alvina, on her way to nursing training, has decided Miss Frost's death in her own heart. Like a novelist musing to herself about her narrative strategy, she thinks: "It was time now for Miss Frost to die. It was time for that perfected flower to be gathered to immortality. A lovely *immortel*. But an obstruction to other … blossoms which were in bud on the stem" (*LG* 36). The mother figure refuses to die, however, and she inhabits the body of someone else. In this case, the revenant mother appears in the form of the exotic carnival woman, Mme Rochard.

Mrs Houghton dies of heart trouble and Miss Frost of pneumonia. Chest troubles continue to plague the characters of this work, for Mme Rochard of the Natchas also suffers from a chest cold. When James keels over in the box office of his cinema, he is ill with "[p]neumonia and pleuritis and a touch of meningitis" (*LG* 173). The chest was, of course, Lawrence's own site of weakness. But instead of a contrast between the sickness of England and the health of Italy that the narrative leads us to expect, Alvina, who has always perked up whenever she has chosen the direction of vitality in the past, wilts under the conditions in Italy. In the unhygienic setting of Ciccio's Italian home, Alvina catches a bad cold as well: "But her chest was raw, she felt weak and miserable" (*LG* 318). What is happening here is that the narrative strategy of the text is being undermined by Lawrence's anxieties. The "killing off" of the characters who have blocked Alvina's flight to fulfillment is ironically counterpointed by her own metaphorical death, which occurs through Ciccio's lovemaking: "And he killed her. He simply took her and assassinated her" (*LG* 202). Kinkead-Weekes writes that anal sex may be implied in this scene and elsewhere in the relationship between Alvina and Ciccio and that "Alvina's 'upper' selfhood is so resistant that it requires to be 'killed' by the dark man" (576). It is possible to argue, as Virginia Hyde does, that "[i]n this process of annihilation, Lawrence is not attempting to deemphasize her essential self but only her socially constructed one" ("Versions of Persephone" 108). But the imagery of confinement and illness as well as the more literal "killing off" of characters in the novel should alert us to something gone amiss either in the narrative strategy itself or in Lawrence's conception of exactly what it is he wants to do. He cannot with any semblance of consistency suggest that those aspects of Alvina that are deathly to her are purged through the deaths of her Woodhouse family and in the "assassination" scene with Ciccio, which also uses tropes of poison and suffocation (*LG* 203), and at the same time insist that the heroine, thus purified, is still too diseased to survive. His own experience in the frigid Abruzzi mountains for a few days in December 1919 may be forging an identification with his chest-plagued heroine that consorts awkwardly with his defensive aggression against her. Her pregnancy, moreover, epitomizes the contradiction in both suggesting her renewed connection to life forces and triggering the nexus of anxiety and resentment Lawrence came to associate with maternity and the "devouring" mother.

Film, finally, is the most unexpected of the symptoms of cultural decay in *The Lost Girl*. Lawrence rails almost hysterically against the cinema in this novel. As with James's hotel, the cinema endeavour has overtones of sexual sublimation. The movie theatre is an "ox-blood red erection" (*LG* 108). A born, if bumbling, entrepreneur if ever there was one, James clashes with Miss Pinnegar over the scheme for a film theatre. As in *The Fox*, the confrontation takes place over a meal. There is both social comedy here and an indication, through the use of the motif of food, that this is a fight about survival, both in terms of the power struggle between James and Miss Pinnegar and more broadly in terms of the survival or demise of a way of life as Lawrence sees it.

At the most basic level, Lawrence objects to the cinema because he views it as the result of the same horrid mechanical principle so brilliantly analysed in the workings of the Crich mines in *Women in Love*. It is an agent of narcissism, one that allows the self to remain within a safe and closed loop of experience. It represents, in reality, the essence of Lawrence's objections to modernity *in toto*. Alvina tries to explain to Mr May why the common people prefer the films to the "turns": "'They hate anything that isn't themselves. And that's why they like pictures. It's all themselves to them, all the time'" (*LG* 116). Madame realizes that film means the eventual end to her career and to her profession. As she explains to Mr May, modern audiences want to experience easily, through the eye, rather than to feel the way the Natchas force them to feel because they are real people (*LG* 149). Movement and growth are marks of livingness. Film presents the *illusion* of life through moving picture, a simulated organic movement that is a sham of life. It appears to be a spatial form, but in reality is temporal. Cinema, then, is a symptom of a people who resist the life-enhancing experiences of otherness and blood knowledge in favour of sameness and ocular titillation. They are a people on the way to a living death. Tony Pinkney writes (in connection with Australia) that mass culture for Lawrence had no essence and reduced "all it touches to one-dimensional surface and ephemeral moment" (114). Film, the prototypical mass cultural medium, had precisely such characteristics for Lawrence.

According to James C. Cowan, who has written the most extended consideration of Lawrence's opposition to film in his book *D.H. Lawrence and the Trembling Balance*, films in *The Lost Girl* represent a "neurotic withdrawal into an antilife convention of projective fantasy" (27). The "imbalance" inherent in fixed pictures that seem to move results in "the immaturity of film heroines, the stereotypical standards of beauty and sexuality in the movies, the encouragement by film of the society's preference for the counterfeit to the real, and the mechanization of consciousness" (27). While all of this rings true and helps us to understand the role of film in the structure of *The Lost Girl*, none of it fully accounts for the extremely hostile view of cinema expressed in the novel, the same shrillness seen occasionally elsewhere in Lawrence, as in the "Nightmare" episode in *Kangaroo*. Here I think we have to look at where Lawrence feels most threatened in a personal way. James Krasner's point regarding Lawrence's view of the relationship between matter and energy may be relevant. One of Lawrence's characteristic traits as a writer is his ability to see beyond the materiality of an object or a person to, as it were, the moving molecules beneath the surface. As Krasner suggests, "Like Darwin, Lawrence makes form fluid, calling into question the stability of the material world, but where Darwin's fluidity expresses the variety of actual biological forms Lawrence's reflects the precariousness of any form" (157). Perhaps the hatred of cinema expressed in *The Lost Girl* can best be explained as the fear of the dissolution of the self displaced onto a form that makes too obvious the instability of the material world and thus of the self.

As in many of the works of the same period, there seems to be a hidden agenda in *The Lost Girl*, hidden in part even from Lawrence himself. While the symptoms of the decadence of English society are manifested in a pervasive and often comical gender confusion and in a cultural embrace of death and dead forms brought out in the popularity of the cinema and in the pervasiveness of illness, the implications of such characteristics are not completely what Lawrence wishes to suggest, as witnessed by several apparently unintentional ironies. For example, Lawrence clearly sets up an oedipal situation in the triangle formed by Alvina, Mme Rochard, and Ciccio comparable to the triangle in *Sons and Lovers* involving Miriam, Mrs Morel, and Paul.

Yet, by the end, there are strong indications that Lawrence's unresolved pre-oedipal problems intrude upon this pattern as Alvina, the wife, begins to look more and more like Mme Rochard, the mother-substitute. One indication of such a development is that the chest ailment Alvina suffers towards the end of the novel harkens back to Madame's illness, the chest pains that first brought Alvina as a trained nurse into close contact with the Natcha-Kee-Tawara troupe, "a parody of a matriarchal family and a carnivalesque 'alternative' to bourgeois patriarchy" (G.M. Hyde 82). Madame also instructs Alvina in cooking and sewing, domesticating her, in effect, in a process similar to Dionys's insistence in *The Ladybird* that Daphne sew shirts for him, and thereby creating a substitute for herself in the relationship with Ciccio (who has lost both his parents). As Ruderman suggests, *The Lost Girl* rehearses in fiction what *Fantasia of the Unconscious* would soon proclaim in prose, "that the mother-dominated household is the cause of neuroses from the smallest to the largest levels of society" (38).

Then there is the framing irony of *The Lost Girl*. Alvina leaves what Lawrence clearly depicts as a confining and deadly England, the "corpse-grey cliffs" seen from the boat and the "long, ash-grey coffin slowly submerging" (*LG* 294). As she leaves England with Ciccio, the feeling of confinement eases palpably and gives way to a feeling of expansion: "And it all seemed bigger, nobler than England. She felt vaster influences spreading around" (*LG* 297). However, Alvina ends up in a primitive home in Italy that is scarcely more commodious and hardly more life-giving than the England she left behind. Furthermore, as Herring writes, "[f]or Alvina … the price of liberation from English provincialism will be paid in being subjugated to an even more rigorous set of conventions in Italy" (13). As with the other transitional works, then, the superficial storyline of the liberation of the Sleeping Beauty from various forms of confinement is disrupted by the return of what looks very much like confinement and constriction at the end. The attempt to break out ultimately fails because of anxiety related to survival.

In addition, Lawrence seems to involve himself in a contradiction he cannot get out of, similar to the one he sets up in *The Plumed Serpent*. He has chosen a female protagonist with whom he clearly identifies to some extent. Lawrence does not want to be another realist writer like Bennett, writing about average humanity: he has a Nietzschean belief that extraordinary individuals move the human race along through creative evolution. But this belief conflicts with his gender politics where the man, any man, must lead. There is even a note of threat that creeps into the text and is in keeping with the deaths of Banford and Mrs Hepburn in the novellas. Alvina feels that "in some way Pancrazio had killed his wife – not consciously, but unconsciously, as Ciccio might kill *her*" (*LG* 336). The spirit of the place, ominously, is one of "ancient gods who knew the right for human sacrifice" (*LG* 315). And we know, from the later "Woman Who Rode Away," who the victims of such sacrifice are destined to be. Clark maintains that in *The Lost Girl*, as in the other pilgrimage novels, "[w]hatever Lawrence's insistence in the American essays that women submit to the control of men, sometimes in raging polemic, the women in the fiction possess a great deal of power and use it skillfully in steering the men in the direction they desire" (199). While it is true that Ciccio (and his uncle) come to depend upon Alvina emotionally, the subversive imagery of illness, awakening, and confinement in the novel and the suggestive placement of Alvina as symbolic mother (as well as literal mother-to-be) all conspire to indicate an unconscious subtextual effort to deprive Alvina of any real power, so that, in the end, we must agree with Roger Fowler's contention that "Alvina's consciousness is subordinated to authorial ideology" (64) and with Peter Balbert's judgment that the leadership fiction is clearly just around the corner in this aspect of *The Lost Girl* (399).

Lawrence's wish to claim and to demonstrate the importance of male ascendency in heterosexual relations and at the same time have a female protagonist is a conundrum. John Worthen suggests that the novel is a "narrator-dominated fiction" written in a way that prevents Alvina's inner life from being our centre of interest (*Idea of Novel* 109*f*). Lawrence's choice of female protagonist, by no means uncommon in his fiction, may be partly motivated by his desire to portray the beauty of a male body and his partly repressed attraction to it. But judging by the novellas, the female protagonist is a choice dictated more by his desire for female passivity to come convincingly from within rather than to be an external demand as in *The Fox*.[18] For, towards the end, Alvina "wished she could lie down there by that town gate, in the sun, and swoon for ever unconscious … There was nothing for her but to yield, yield, yield" (*LG* 321). Lawrence's unsatisfactory solution to his problem, then, seems to be to suggest that self-effacement is the way to self-fulfullment for this essentially *Bildungsroman* protagonist, a suggestion that clearly is nonsense. As Marguerite Beede Howe puts it, "[t]he irony of *The Lost Girl* is that Alvina suffers the anguish of dissolving personality in order to extinguish her social self" (25).

And yet, there is a hint of something different at the end of the novel, endings being a point where Lawrence often tends to make a courageous attempt to be ruthlessly honest with himself and step back from his obsessive fantasies. The ending, in which Ciccio is called up for military duty in the war and leaves Alvina with a promise to return and travel with her to America, has usually been interpreted as an optimistic if open conclusion to the novel. Although we are free to imagine almost any eventual outcome,[19] it may well be that it is Alvina's self and not Ciccio's that will survive finally. But Alvina warns Ciccio that "'[i]f you don't come back it will be because you don't want to'" (*LG* 339). We have to consider the distinct possibility that, in the Lawrentian scheme of things, it is unlikely that Ciccio will come back. For Alvina is about to be a real and not merely a symbolic mother, and mothers for Lawrence, especially in this period of writing, are deadly. The baby is there ready to displace Ciccio in his relation to Alvina just as Dr Mitchell's imagined baby would displace him (and as Frieda's children, Lawrence felt, tended to eclipse him in Frieda's priorities).The war will do its culture's dirty work.

I have suggested how the word "rainbow" is a key term that allows Lawrence to look back on his earlier fiction and criticize, even parody, it, particularly in *The Fox* and *The Lost Girl*. Both Diane S. Bonds and Anne Fernihough have studied the ways in which various other Lawrentian terms morph in the fiction to the point where, in some cases, by the end of a work the word suggests the opposite of what it did at the beginning.[20] One word that Bonds examines closely is the word "surety" in *The Rainbow*. In *The Lost Girl*, the final spoken word – "Sure?" – functions as a poignant commentary on the loss of the surety that was emphasized throughout, but especially in the second paragraph of, *The Rainbow*: "There was a look in the eyes of the Brangwens as if they were expecting something unknown, about which they were eager. They had that air of readiness for what would come to them, a kind of surety, an expectancy, the look of an inheritor" (*R* 9). After the war, that unknown has arrived, and it is neither a pretty sight nor an event that any amount of readiness could have prepared one for. The loss of surety, unfortunately, resides not only with the characters Lawrence has been describing but also, metafictionally, in his own writing. What the reader inherits is bad fiction.

AARON'S ROD: FLOWERS AND TOWERS TO COUNTER FEMALE POWERS

In *Aaron's Rod*, we are a step beyond the world of *The Captain's Doll*, where marriage to a younger wife must be accompanied by a redefinition of love, and the world of *The Lost Girl*, whose unusual history of composition ensured that traces of the imaginative universe of *The Rainbow* remained detectable. Now marriage itself, the angelic ideal of *The Rainbow*'s Tom Brangwen, is the focus of denunciation, although even such an attitude obliges the text to keep one foot in the old world, if only the better to kick it. *Aaron's Rod* seems to repudiate *The Lost Girl* in some ways. The paradox of losing oneself to find oneself is here put into the discredited mouth of James Argyle (a character based on Norman Douglas): "'What could be better than to lose oneself with someone you love, entirely, and so find yourself,'" he asks rhetorically (*AR* 240). Once again, Lawrence seems to be engaged in a dialogue with himself or with his own fiction and to be rejecting a previous stand. Where *The Lost Girl* attempts to present a domestic "solution" to the problem of confinement, *Aaron's Rod* begins with a domestic situation and proceeds toward a public resolution that really resolves nothing. Lawrence's writing at this point was somewhat obsessive and formulaic because he was deliberately reviewing his old stance in order to get beyond it. Such writing usually means that, against his own advice in *Studies in Classic American Literature*, Lawrence was trying compulsively to force a "metaphysic" upon his fiction. That attempt, in turn, usually means that there will be a disjunction between the surface and underlying codes.[21] Lawrence believed that, in *Aaron's Rod*, he was writing a novel about a man who feels confined in marriage and about his development toward independent "proud singleness." It demonstrates the necessity of such singleness and the preferability of male friendship to heterosexual love. The positive spin one can impute to such a development is that "the novel seeks the reinvention of relation" in general (Vine xxiv), but the energy of the fiction is invested much more in the fight against the connection with women than in an improved relation with anyone. Severe survival anxiety results in subversive imagery – particularly of breathing and cannibalism – that suggests that what lies behind the leadership ideal of male ascendancy is a defensive reaction to the reality of male dependence and the fear of the "devouring" mother that goes back to the pre-oedipal period.

Aaron Sisson, the novel's protagonist, is a relatively well-educated and musical miners' check-weighman who decides to abandon his wife, Lottie, and their two young daughters for no apparent reason. He leaves his Midlands home without a word of farewell, and soon finds himself taken up as a kind of class curiosity by a group of revolution-spouting bohemians, at the centre of which is Jim Bricknell (a character based on a Captain James White)[22] and at the fringes of which we find the Lawrence persona, Rawdon Lilly. In London, Aaron finds work as a flautist in an orchestra and has a brief affair with one of the bohemians, Josephine Ford (Lawrence's portrait of Dorothy Yorke). Restless and unfulfilled, he sets off for Italy in search of Lilly and stays first in Novara with the Franks and then in Florence, where he becomes entangled with an American expatriate, the Marchesa Del Torre (Tower). When he is finally reunited with Lilly at a café in Florence, Aaron sees his flute destroyed in an anarchist explosion that rips through the establishment. The novel ends on a tentative "note," with Aaron experiencing a cryptic dream and Lilly urging him to submit to the greater soul in Lilly himself.

Under its rather aimless picaresque veneer, *Aaron's Rod* is full of contradictions and leaps of logic. It consistently blames women for a series of failed heterosexual relationships in which a man, Aaron, is the common denominator, and it illogically associates the war with love rather than with power. Nevertheless, various close readings of *Aaron's Rod* in recent years have clarified its rather amorphous structure and its system of allusions.[23] Steven Vine, for example, is helpful in relating the text to its historical context, a connection that allows us to see its uncertainty as "an image of the historical, sexual and psychical crises of the post-First World War period as Lawrence saw them" (xv). As Vine suggests, the war has made the Lawrentian quest for selfhood, in particular the masculine definition of selfhood, both urgent and ironic in *Aaron's Rod*. It is urgent in that the war "engenders … a bizarre feminization of the male characters" and ironic in that the novel "refutes the idea that the self is a fixed or a given entity" at all (Vine xxviii, xxvi). Surprisingly, the very problem that lends urgency to the quest becomes, in effect, the solution in that *Aaron's Rod* "relocates femininity within the body of the masculine" (xxx). Although Vine at times recognizes the self-reflexive features of the novel, what is missing in his very good essay, I feel, is a taking into account of Lawrence's own life and anxieties, which, as we know, revolved closely around the issue of gender and gender relations. Vine points out that "Aaron comes to consciousness largely through Lilly's denunciations of love, sociality and modernity" (xxv), and countless readers have recognized that Lilly and Aaron represent two aspects of Lawrence himself or at least a splitting of the perceiving consciousness in the novel. A central paradox in the work, then, is that the theme of singleness is developed through a central figure who is divided into two. The two sides of the character are, however, unequal. There is a privileging of Lilly, who is introduced to us as a writer – one of the few writers in Lawrence's fiction – and who might be seen as the author himself inscribed in the text. The fictional character Aaron has the ability, like some existentially free comic-strip figure, to enter his creator's drafting room and ask for an explanation for the goings-on in his comic-strip life, while Lilly gives voice to Aaron's feelings, as any good writer will do for his fictional creation.[24]

Lawrence is not only interested in the fate of his picaresque hero, Aaron, but also keenly engaged in the survivability of the Lawrence figure, Lilly, under the pressure of a fictionally enacted clash of ideas. Lilly, largely an embodiment of a system of ideas, and even a windbag at times, represents for Lawrence his own efforts as a writer to evolve something new by persuading Aaron to embark on a new course. For both figures, there can be no looking back to the dead forms of the past. For Aaron, that means making a break with married life, and for Lilly/Lawrence it means saying good-bye to the marriage novels. But there is more to it than that.

In "She Looks Back," a remarkable poem from *Look! We Have Come Through!* (1917), Lawrence expresses his distress and bitterness at Frieda's hankering after her children in England in terms that associate her salt tears with Lot's wife, who was turned into a pillar of salt when she looked back at the destruction of Sodom and Gomorrah in violation of a divine interdiction. In the poem, the punishment is not only the wife's, for the speaker feels "[b]urning of powerful salt, burning, eating through my defenceless nakedness" (*Poems* 207). The bitter tears and his bitter feelings expand into a denunciation of motherhood itself: "And that mother-love like a demon drew you from me / … The mother in you, fierce as a murderess … / Not Wife, but Mother. / I have learned to curse your motherhood" (*Poems* 207). The very end of the poem strongly implies that the fear and hatred of motherhood derives not, in the first instance, from marriage but from childhood, a memory that marriage has served to reawaken. The curse against "[a]ll mothers who fortify themselves in motherhood … burns within me like a deep, old burn" (*Poems* 208).

The "pillar of salt" metaphor plays a central role in the novel and appears in two chapter titles. It implies the need for a man to go forward away from marriage most obviously, but, as we have seen, Lawrence also associated it with his grudge against motherhood in his own life, and that feeling becomes part of the complex importation into the fiction of extra-fictional concerns. If *The Lost Girl* hints at the pre-oedipal obsessions that are at the foundation of Lawrence's survival anxiety, *Aaron's Rod* magnifies them to clear visibility. When, at the beginning of the novel, Aaron's daughter Millicent breaks a Christmas ornament, a blue ball, the narrator's words of free indirect discourse that describe Aaron's thoughts – "And this was the end of it" (*AR* 11) – prefigure Lilly's words to Aaron at the end of the novel regarding his smashed flute: "'Throw it in the river … It's an end'" (*AR* 284). The novel comes full circle from one scene that is an emblem of the end of a marriage to one that implies, in the "broken rod," the end to all heterosexual activity for Aaron. But in the first scene, Aaron momentarily acts more like a sibling to his daughter than a father, while Aaron's wife is in the role of the mother mediating (not very constructively) between two quarrelling children: "'She wanted to break it,' said the father. 'No she didn't! What do you say that for!' said the mother" (*AR* 11). The breaking of the blue ball, Aaron's childhood possession, may possibly be Lawrence's signal that Aaron's childhood dependency is at an end, but, in light of Aaron's subsequent relations, the more persuasive reading of the event is that Aaron's infantile needs have been frustrated and that he is determined to have them fulfilled elsewhere. While Lawrence is embarked on a new course for his fiction that will displace marriage as the centre of concern, his own still troubling pre-oedipal deficits continue to be played out at a subtextual level.

The pillar-of-salt chapters in *Aaron's Rod*, moreover, may be said to function as something of a self-reflexive signature. Some of Lawrence's works of fiction contain a scene that seems to enact the very form of the work itself.[25] In *Aaron's Rod*, Aaron, to confuse the biblical allegory, returns home and voyeuristically watches his wife inside the house from the garden. The "Pillar of Salt" return (the title of chapter 4) is repeated in "More Pillar of Salt," chapter 11, when Aaron again visits his home and this time makes contact. Aaron, in effect, is looking back to marriage before moving on to the male bonding in the leadership culture. The scenes are fictionally significant in suggesting Aaron's difficulty in casting himself loose from the marriage union and, as we can see, thereby cutting the umbilical connection to a mother figure. They are also metafictionally meaningful in characterizing the revisionist tendency of the transitional period of writing itself and Lawrence's own self-conscious attempts to travel imaginatively from marriage to leadership while warning himself of the dangers of looking back. There must be no going back to the twoness of marriage for the male seeking the oneness of self-fulfillment. Only strong boundaries around the single self are adequate for individual male survival, given the potential "murderess" in every woman/mother, but such fortress-building brings about problems of its own. The main problem is the difficulty of maintaining a delicate balance between the need for walls of protection and the fear of confinement associated with female power that those walls evoke. Aaron's quest, boiled down to the abstract essence that embodies Lawrence's own dilemma, becomes a search for a balance between the need for protection and the fear of a smothering/mothering confinement.

The horrors of confinement are consistently associated in *Aaron's Rod* with heterosexual relationships, in particular as they impinge upon Aaron. On the street where Aaron has his home, there are "[s]o many houses cheek by jowl, so many squirming lives" (*AR* 39). But even the pub, his normal refuge from domestic asphyxiation, seems confining with its landlady in her "tiny parlour" (*AR* 17). On the day he deserts his family, Aaron leaves the pub and finds himself outside intending to go home, but "[t]here seemed a wall in front of him. He veered. But neither could he take a stride in the opposite direction" (*AR* 25), so he spins round and round like the coiling of the spring that eventually sets the centrifugal force of his quest in motion. When he looks back upon his marriage later in the book, Aaron imagines that his wife's female will is "diabolical" in that it "could press like a flat sheet of iron against a man all the time" (*AR* 158) and, in something of a delayed mixed metaphor, that it is like "a flat cold snake coiled round his soul and squeezing him to death" (*AR* 161).

London is not much better in creating a feeling of freedom for Aaron. The opera attended by Josephine, Jim, and company, during which they spy Aaron in the orchestra, is *Aïda*, a work that culminates in the entombment of the heroine and hero. The opera box itself is a "narrow space" (*AR* 52) in which the spectators feel like marooned "Robinson Crusoes" (48), and the party (including Lilly and Aaron) proceed from that confining space to the underground "tube," down "in the bowels of London" (*AR* 63). When Aaron's own bowels cease to function after his love bout with Josephine, we have yet another manifestation of something trapped within a restricted space. Until his constipation is relieved through the maternal ministrations of Lilly, Aaron remains dangerously ill both physically and psychologically,[26] for the psychological problem behind this literal anal retention is the fear of the loss of self.

Having got England out of his system, so to speak, Aaron journeys to Italy, where he seems to be getting on quite well until, at the home of Sir William and Lady Franks in Novara, he cannot help feeling like a caged animal: "He felt himself dashing up the steps into the garden like a bird dashing out of a trap where it has been caught" (*AR* 150). This is in spite of the hospitality or, rather, because of it, since patronage in some sense re-enacts the parent-child relationship that Lawrence believes women have transferred to modern marriage. Even the logs in the fire seem to Aaron to be "like leopards slumbering and lifting their heads within their cage" (*AR* 154). The landscape on view during the railway journey to Florence seems to promise a modicum of liberation; it is so different from England: "No more the cosy English ambushed life: no longer the cosy littleness of the landscape" (*AR* 199). Likewise, initially, Aaron's affair with the Marchesa: "And he felt all the bonds that held him melting away" (*AR* 251). Apparently a slow learner, Aaron forgets that, in this Lawrentian world, opening up can allow the entry of dangerous threats to self-integrity: "'I gave myself away,'" Aaron thinks to himself (*AR* 230). No sooner does Aaron go to bed with the Marchesa than he feels trapped: "He wanted to get out of her arms and her clinging and her tangle of hair" (*AR* 262). Easier said than done, though, for during a nightmarish few moments, Aaron feels literally locked in within the Marchesa's home as he fumbles with latches and struggles with locks (*AR* 274); suddenly "he felt that the doors of Florence were trying to prevent his egress" (*AR* 275).

According to Tony Pinkney, it is possible to trace the misogynist association in Lawrence of male confinement with women at least as far back as *Women in Love*, where a reworking in the womb imagery of *The Rainbow* signals a change in attitude: "Wombs in the later novel are an intolerable claustrophobia, the prison-bars of the old world rather than an opening to the new" (86).[27] Gerald Crich, we may recall, has an extreme fear of being bound. With regard to *Aaron's Rod*, Pinkney sees a rather bizarre identification of the war with "a voracious, absorptive vagina enticing a reluctant but powerless male into its soft, deadly depths" (118). This may be an extreme interpretation, but certainly, for a few years after *The Rainbow*, Lawrence's threshold imagery of holes in walls, arches, doorways, etc. gives way to images of confinement in keeping with how Lawrence himself viewed his life, prevented as he was from leaving England during wartime. In *The Lost Girl*, the deaths of Mrs Houghton, Miss Frost, and, finally, Mr Houghton progressively free Alvina from the bonds that keep her confined within the conventions of Woodhouse. In *Aaron's Rod*, death is not required for the survival of the protagonist, at least not the death of recognizable characters in the fiction, for where *The Lost Girl* ends on the edge of the precipice of the First World War, *Aaron's Rod* shows its aftermath, when death has become generalized in the culture. As well, the war has destroyed, in Aaron at least, all restraint based on social convention. In what seems to be a harking back to *The Rainbow* with a difference, there is a death within Aaron that the text attempts to portray as positive. He must be reborn, but rebirth carries with it the danger of a return to infancy in symbolic form. Such a development in the postwar Lawrence is fraught with risk.

That survival is what is at stake in *Aaron's Rod* is indicated for one thing by a scene late in the book. Aaron, in recoil from the Marchesa, seeks restoration in a solitary communion with the cypress trees in the Tuscan countryside. Ghosts are mentioned, as are "lost races, lost language, lost human ways of feeling and knowing," and respiration, the most basic activity needed for survival by land animals, is emphasized: "He lay and watched tall cypresses breathing and communicating, faintly moving and as it were walking in the small wind." The cypresses seem "like so many high visitants from an old, lost, lost subtle world" (*AR* 265). In a poem entitled "Cypresses" (first published in 1923), evidently inspired by the same landscape portrayed in *Aaron's Rod* and also full of a dark realization of dead species, particularly the dead race of the Etruscans, there is a closing meditation on survival: "They say the fit survive, / But I invoke the spirits of the lost. / Those that have not survived, the darkly lost" (*Poems* 298). In *Aaron's Rod*, at this point, as in *Women in Love* at times, evolution seems to offer a hope: "That there has been and will be life, human life such as we do not begin to conceive" (*AR* 265). But this Darwinian development can be contemplated only at intervals when the hold of the self seems to dissolve temporarily, as here for Aaron when "his soul seemed to leave him" (*AR* 265). Usually, such a merging is felt as a threat to the survival of a fragile self. In fact, Aaron emphatically rejects his own former belief that "the consummation of human life ... was this flinging oneself over the precipice, down the bottomless pit of love" (*AR* 165) – similar to the "bottomless gulf" of the attempt to find happiness in *The Fox* (*Fox* 69) – and he refuses to step "over the edge of the abyss of selflessness" (*AR* 165).

The fight in *Aaron's Rod* against the perceived priority of woman is also a fight for the survival of the male self, which in this text is equatable for the most part with individuality. Women are set up as the advocates of the sublimation of individuality for the common good. The landlady at Aaron's favourite pub, the Royal Oak, argues against the "selfish policy" of men (behaviour that, presumably, includes their patronizing her establishment as a refuge from home life) and in favour of "'improving the world you live in'" (*AR* 21). This is a sophisticated echo of Lottie's complaint against Aaron's selfishness. Lilly's soliloquy against the "*bonum publicum*" in "Low-water Mark" counters the female view. As Michael L. Ross has pointed out, George Eliot's quest in *Romola* for "a model of community" in Florence is reversed in the Florence section of *Aaron's Rod* (*Storied* 91). John Edge's argument that power in the leadership period "demands the submersion of the individual in the collective" (43), as far as this novel is concerned, mistakes the female position for the text's "official" one. In fact, Edge's claim must be turned on its head to make sense in the context of *Aaron's Rod*. Power, according to Lilly/Lawrence, is necessary to combat the love ethic that destroys selfhood (male) in the name of the collective (female).

The boundaries protecting the self are imagined in *Aaron's Rod* in unusually concrete terms that are related to Lawrence's conscious rejection of Darwinian concepts. Lawrence did not share the modernist obsession with time, an obsession that owed its birth to the intellectual ferment that initially culminated in Darwin's work and whose later source was Henri Bergson's influential assertion that our habitual thinking in spatial terms is the cause of our speciously attributing materialist explanations to human consciousness (Levenson 41). Lawrence argued adamantly against the exclusive cultural foregrounding of time (which he attributed to Darwin, Spencer, and Huxley). In the Hardy study, for example, he insists that spatial constructs are at least as crucial to understanding the world around us as temporal ones (*P* 485). His portrayal of the attempt to reconfigure and strengthen the sense of selfhood in *Aaron's Rod* depends not on the temporal deepening of memory (as in Woolf's *Mrs Dalloway*) but on the spatial widening of a feeling of connectedness between the self and the world around it. The work of this period of Lawrence's writing is a reaction to the same conditions that other modernists saw as threats to selfhood, but Lawrence detested the pessimism and tragic vision of some of his contemporaries. The hollow self of T.S. Eliot or Joseph Conrad does not result in Lawrence in a spatialization of time. Where most of Lawrence's contemporaries portrayed the reduced potential for selfhood in modern life through the "flattening" effect of spatialization and their characters' attempts to combat such straitened conditions through the "depth" of memory,[28] Lawrence, although certainly capable of creating vivid portraits of hollow men such as Gerald Crich in *Women in Love*, attempted, in *Aaron's Rod* preeminently, to use spatial forms assertively as images of renewed selfhood even while constructing a conventional plot with "normal" time sequencing. The result almost look like an inversion of the typical modernist method of dealing with the spatial-temporal dimension.

Aaron's Rod is a ploddingly chronological work that uses spatial forms to explore the potential for male self-actualization in the face of what is portrayed as a female threat to "flatten" male pride. The spatial forms are images of both the necessary protective covering of the self and the confining forms that threaten it. Even as Aaron's journey over the months is carefully charted, beginning with Christmas and ending the following November, the emphasis is not on development over time but rather on a growing "spatial" awareness of the "real" (male) self and how the self faces major hurdles to actualization in the modern world, particularly within the structure of modern marriage. Lawrence here uses the ideal of the male self as a rooted plant to fortify the vulnerable masculine ego. Lilly wants to see men get back up to a vertical position, standing on their legs once more in proud defiance of woman, who has "climbed up with her children"

(*AR* 101) to a position of "institutionalized Mariolatry" (Virginia Hyde, "Revisionist Typology" 477). Thus Lilly is disgusted at Jim's inability to walk properly (*AR* 82) and alarmed when he sees Aaron, seemingly drunk, fall on the pavement below his room in London (*AR* 87). Only the firmly rooted self can withstand the buffeting of modern life and modern women. The ideal is symbolized in the flower of Tuscany and its upright creations: "Florence, the flowery town ... Flowers with good roots in the mud and muck, as should be: and fearless blossoms in air, like the cathedral and the tower and the David" (*AR* 232).[29] Flowers and towers will counter female powers! Lilly's name suggests something similar, of course. In keeping with Lawrence's characteristic use of botanical images as touchstones for human existence, the ideal male self in *Aaron's Rod* is secure in its place as a rooted plant is, and it is Lawrence's favourite image, the tree, that most centrally represents ideal self-sufficiency. However, Western culture creates an environment that is not always conducive to that condition.

The Christmas tree in the Sisson home with which the novel begins has been severed from its roots, but even the growing tree in the Bricknell garden of the third chapter, "The Lighted Tree," (to which Aaron comes from the Royal Oak) is ironic. As Virginia Hyde argues in her study of the biblical typology of the novel, neither tree "embodies the flowering rod of nativity or the flowering cross of resurrection" for its worshippers are parodies of genuine worshippers just as Jim Bricknell's compulsive eating of bread and drinking of wine is a burlesque of the sacrament ("Revisionist Typology" 475, 476). And yet Aaron, in watching the revellers, is "rooted" (*AR* 34) despite the fact that he is on his spring-like "recoil" away from Lottie at this point. The ideal state of a rooted and flowering plant is apparently difficult to achieve. Aaron is likened to a chestnut that has "tumbled from the tree of modern knowledge, and cracked and rolled out from the shell of the preconceived idea of himself" (*AR* 164). But it is doubtful that he has taken root as yet or that he is about to flower.

Modern systems are particularly hostile to self-development in the postwar period, and spatial voids in *Aaron's Rod* are critical in indicating the crisis of identity. Paul G. Baker argues that the novel's recurring images of abyss, emptiness, and hollowness reflect conditions in postwar Europe (109). The world of the novel is one in which the modernist void in values results in near Conradian hollow men and women. Money, it seems, is "the only authority left" in a world where "[t]he feeling of emptiness, of neglect, of lack of supplies was evident everywhere" (*AR* 136, 152). Even the opera house in London is a "hollow shell" (*AR* 45). In Milan, where Aaron witnesses a mob scene, "there seemed a curious vacancy in the city – something empty and depressing" (*AR* 180). Many of the people who inhabit this world are void within. Jim Bricknell feels that he must continually stuff himself during all hours of the day or night in order to fill the vacuum left by the diminishment of his sense of self. Josephine is plagued by an existential angst that results from her feeling of nullity: "'Sometimes I think I'm nothing when I'm alone. Sometimes I think I surely must be nothing'" (*AR* 68). Julia has never given a thought to her own self (*AR* 48). Even "our hero," Aaron Sisson, finds himself in a sort of vacuum. Once he has broken "the bonds and ligatures which bound him to the life that had formed him …, [t]here was nothingness. There was just himself, and blank nothingness" (*AR* 178). Aaron's host in Novara, Sir William Franks, has been awarded three medals for his war efforts, one of which is pinned over the pit of his stomach. This is not to recognize his "greedy appetite" as he jokingly suggests (*AR* 140), but rather, in the spatial scheme of the novel, to balance a hollowness within with what the narrator sees as a vacuous gesture without, represented in the flimsy pieces of metal. While Jim believes that only love (or bread) can prevent the empty feeling at the pit of his stomach, Aaron finds that love empties him to such an extent that he weakens and falls ill.

Bearing in mind Marguerite Beede Howe's assertion that each of Lawrence's novels "represents an attempt to reach an equilibrium with reality" (2), one can say that in *Aaron's Rod* inner and outer pressure must be in equilibrium to avoid implosions into the void or explosions out to it. The pressure from the outside comes from feelings of confinement. (The peripatetic nature of Aaron's quest is determined in great part by his need to avoid feeling confined.) But the inner pressure that must be there to keep the outer force from crushing the self is missing. There is a void inside, and Aaron's quest for selfhood can be said to be a "spatial" struggle to fill it. If the vacuum created by the void within society and individuals is not filled by something more constructive than bread, the mob and the bomb will result. Imbalance can also occur through outer pressure increasing to the breaking point, rather than inner pressure weakening. Such is the case when Jim punches Lilly in the solar plexus, thereby creating a void in Lilly's middle that makes him resemble Jim and Sir William, however temporarily. It takes all of Lilly's concentration not to lose control over his breathing as air rushes back into his lungs to fill the void. Similarly, in Aaron's case, it has been the external pressure of his marriage that has created the feeling of disequilibrium and prompted him to flee. But since he is at least one half of the problem in his marriage, the feeling of confinement he is attempting to escape will follow him wherever he goes until he strengthens the self within to the point where it can balance the pressure from without. Lawrence develops an elaborate set of images to explore the balance between inner and outer pressure. Using breathing and explosion motifs, he links the state of the individual to that of a society very much in flux.

Fairly early on in the novel, when Aaron tries to formulate to the Bricknell set why he has left his wife, the best he can come up with is "'I wanted to have some free room round me – to loose myself ... I wanted fresh air'" (*AR* 66). Confinement and lack of air are subtly connected in this work in numerous ways. Breathing begins at birth, at the point of initial differentiation between mother and child. At home, Aaron feels stifled: "[t]he little room was hot," and so he opens a ventilator "letting in a stream of cold air, which was grateful to him" (*AR* 12). Similarly, in Florence, Aaron feels glad to be away from the cosy warmth of his English home: "it had stifled him till he felt his lungs would burst" (*AR* 209). At the pub, Aaron sees in the "void, glistening eyes" of an Indian doctor "the same menace that he saw in the landlady" (*AR* 24, 25), a benevolence that to him is like "poison gas" (*AR* 25). Lilly mocks Jim's contention that "[l]ove is the soul's respiration," while, when the soul exhales, it is "a bloody revolution" (*AR* 62). When Lilly mocks Jim's view of love in "A Punch in the Wind," of course, Jim forces Lilly to exhale. In Aaron's illness, his bowel problem is accompanied by "troubled breathing" (*AR* 90).

Lawrence struggled throughout his life and writing career with boundary definitions in an effort to strengthen his own sense of selfhood, which had been badly damaged by mothering that tended to suffocate that sense and to violate boundaries, as the title to his largely autobiographical *Sons and Lovers* indicates. Lawrence even seems to blame his mother indirectly for his consumption in *Fantasia of the Unconscious*: "On the upper plane, the lungs and heart are controlled from the cardiac plane and the thoracic ganglion. Any excess in the sympathetic mode from the upper centres tends to burn the lungs with oxygen, weaken them with stress, and cause consumption. So it is just criminal to make a child too loving" (FU 59). A similar fear of suffocation is evident in racist remarks by Lawrence while in what was then Ceylon, en route to Australia. There is something approaching hysteria here in the need for hierarchy and separation: "Those natives are *back* of us – in the living sense *lower* than we are. But they're going to swarm over us and suffocate us. We are, have been for five centuries, the growing tip. Now we're going to fall. But you don't catch me going back on my whiteness and Englishness and myself" (*L* iv 234).

The larger issue of Lawrence and race is beyond the purview of this study. However, a brief discussion here will be useful in suggesting that it is the same sense of suffocation that Lawrence's vulnerable male protagonists feel vis-à-vis women that actuates a good deal of the racist remarks of the transitional period. Lawrence's well-known portrayal of Loerke in *Women in Love* has strong anti-Semitic overtones, but the anti-Semitic attitude seems to intensify in the works between *Women in Love* and *Kangaroo*. *Aaron's Rod* may not be any worse in this respect than *The Captain's Doll* or *Mr Noon*, but the peculiar quality of the anti-Semitism in *Aaron's Rod* helps to clarify its presence in the period and allows us to better understand the nature of the survival anxiety that undermines the conscious narrative strategy. Lawrence's attitude toward Jews seems to have been determined in part by the historical context within which he lived and worked. There was a linkage, explored by Sander L. Gilman, for example, in nineteenth-century Europe among Jews, women, and illness: "those defined as 'not German' in Germany, such as the Jews … often are labelled as having a body form that reveals their potential for illness. By the nineteenth century, male Jews are seen as feminized, belonging to an 'inferior race' … The visualization of the difference of the male Jew is thus in terms of this image of the unproductive and the ill, specifically the image of the tubercular female, who is simultaneously 'beautiful' (but dangerous) and 'diseased'" (60). It is easy to see how such a collocation of referents in German culture, which Lawrence was very familiar with, might provoke survival anxiety in a tubercular writer with a particularly deep-seated ambivalence toward women.[30]

In *Aaron's Rod*, the inclusion of Jews among those who are to be made slaves immediately in Argyle's harebrained reactionary scheme (a scheme endorsed in large part by Lilly) seems to be racially motivated and thereby to support Jascha Kessler's argument about the differentiation of blood being needed by Lawrence for establishing servile races.[31] Argyle tells Levison, his interlocutor, that he would make slaves of "'[e]verybody, my dear chap: beginning with the idealists and the theorising Jews'" (*AR* 279). However, we must think back to earlier in the book, to the scene where Aaron is at the Royal Oak pub and Lawrence describes his apparently ongoing flirtatious relationship with the landlady there, an obvious foreshadowing of his much more developed liaison with the Marchesa Del Torre. First, the landlady is called "a large, stout, high-coloured woman, with a fine profile, *probably* Jewish" (*AR* 17, my emphasis). Her nose has "a subtle, beautiful Hebraic curve," and she herself is one "who loved intellectual discussion" (*AR* 18). This is important in the light of Argyle's comments regarding "theorising Jews." A few pages later, the narrator has apparently studied the woman's genealogy, for now she is simply "the Hebrew woman," and her lustfulness rather than her intellectualism is foregrounded (*AR* 23). There is an elision here between race consciousness and a threatening, because sexually demanding, femaleness. Theorizing Jews, people Argyle apparently feels a need to control, turn out, in the subtextual scheme of the novel as a whole, to be associated with lustful women who would enslave the male through the arts of seduction.

The underlying fear of suffocation surfaces in *Aaron's Rod* in the form of an interest in the act of breathing, an obvious source of constant anxiety for the consumptive Lawrence. Controlling one's breath is necessary not only for Lilly to save face after Jim's blow to his midsection but for a flautist like Aaron to play music and for someone like the Marchesa to sing. The awakening that Aaron effects in the Marchesa comes about through the sound of his breath through his flute and results in the liberation of her breath in song: "Then her soul and her voice got free, and she sang … Her soul seemed to breathe as a butterfly breathes" (*AR* 256). In *Aaron's Rod,* it seems that the price of the awakened woman's liberation is the confinement and stifling of the man who is the stimulus for her development. It is as if Aaron and the women he encounters are in competition with each other for the role of Sleeping Beauty, as Lawrence's evolving ideology comes up against his successful writing formula. In Novara, Aaron's "old sleepy English nature was startled in its sleep. He felt like a man who knows it is time to wake up" (*AR* 151). However, Manfredi, in effect, cajoles Aaron into the role of a sort of pied piper Prince Charming opposite his wife's Sleeping Beauty: "'I wish something would rouse her up, you know'" (*AR* 224). Before Aaron plays his magic flute, the Marchesa "spoke almost like one in a trance – or a sleepwalker" (*AR* 226), but afterwards and after her own singing, she is "like one just coming awake" (*AR* 256). With Lady Daphne in *The Ladybird,* "there had always been a pressure against her breathing. Now she breathed delicate and exquisite" (*Fox* 213). Aaron has the same salubrious effect on the Marchesa's respiratory system here. The usual ending to the Sleeping Beauty story is not to be, however, for Aaron, wanting to be awake himself, abandons his role in the fairytale drama with the Marchesa.

The music emphasized in the novel is wind-driven, but there is also the plosive art of Lilly's words. In fact, there are chapters headed "Talk" and "Words." The expelling of air in music or talk is a benign manifestation of an explosive force in the novel, a force that becomes violent when disequilibrium reaches a breaking point. In the novel's first paragraph, the relief at the ending of the war is, as Keith Sagar has suggested (106), subtly described in terms close to explosiveness: "A man felt the violence of the nightmare released now into the general air" (*AR* 5). The first obvious inkling of the discharges to come occurs in the opening chapter, when Millicent breaks the fine blue ball of hardened glass by deliberately dropping it on the floor: "it fell with a little splashing explosion: it had smashed" (*AR* 11). The explosion and breaking glass adumbrate the bomb explosion in the café in Florence. There, Aaron emerges from "the fearful blow of the explosion, the noise of glass, … the rushing of men, the sudden gulf, the awful gulfing whirlpool of horror in the social life" (*AR* 282). John Burt Foster, Jr suggests that the bomb explosion, a political event, "has become the symptom of a larger cultural crisis" (248). I would argue, rather, that Lawrence's linking of the blue ball to the café incident suggests a connection, similar to the one in Conrad's *The Secret Agent*, a novel that also features an "anarchist" bomb, between the domestic and political (the private and public) and the void space at the centre of both.[32] The bomb, in fact, may in some way represent the outward manifestation of an anarchism within Aaron, who tells the Marchesa at one point that orchestra music makes him "want to throw bombs" (*AR* 225). But what is most interesting in the description of the café bombing is the use of the words "gulf" and "gulfing," which creates an implicit connection to Aaron's relationship with the Marchesa as described just a few pages earlier: "He did not dislike her. But he would never see her again. A great gulf had opened, leaving him alone on the far side" (*AR* 275). It is not that domesticity *per se* must lead to violence but that the very connection of men to women leads inevitably to a void space between the two people involved and to a fear of engulfment within the male self. The explosives used in wartime, which Captain Herbertson speaks of (*AR* 114), continue to reverberate after the war. In Milan, for instance, the mob "broke as if something had exploded inside it" when it is attacked by the *carabinieri* (*AR* 185). When Aaron reconsiders his life with Lottie, he realizes how he kept a false image of himself intact in order to avoid conflict, but now "[h]is mask, his idea of himself dropped and was broken to bits" (*AR* 163). Like the blue ball earlier and the broken flute later, this image of fragmentation is meant to suggest the end of a significant life phase or attitude. For Lawrence, this potentially explosive relinquishment of the social mask, something Gerald Crich never accomplished, is a *sine qua non* for the development of a genuine selfhood.

Aaron is the most prominent example of a modern individual struggling with pressures from within and without, but he is far from alone in his explosiveness. The Bricknell group is similar. The paintings in the Bricknell house "exploded their colours" (*AR* 27), for instance. Jim himself "rolled in his chair in an explosion of sudden, silent laughter" (*AR* 31). And, perhaps predictably, they believe in revolution in an armchair, distant kind of way. Spatial forms in *Aaron's Rod* become emblems of a deep-seated problem in contemporary life, where, for Lawrence, hierarchies of class are far too rigid, those of gender far too lax. The hollowness of modern life is leading to a political and social crisis of explosive dimensions, but even at his most bitter, Lawrence is never totally devoid of hope (as some of his contemporaries seemed to Lawrence to be). Indeed, for the typical modernists, as Kern puts it, "the void supplies the focus. Their characters seek meaning outside themselves … and find only the horror of nothingness within" (170). In *Aaron's Rod,* the hope resides largely in what might ensue within the individual and, by extension, in society if people are able to develop balanced and integral selfhood. Spatial forms may symbolize the challenges to such selfhood, but the primary means of envisioning the hope is the portrayal of a specific spatial form, that of "aesthetic space."

One of the most striking passages in *Aaron's Rod* occurs in the dream sequence opening the final chapter. The description of the dream itself is quite long and involves a number of events, in particular a boat ride where Aaron is both passenger and spectator, that have been interpreted variously by the commentators on the book. Near the beginning of the dream, before the boat ride, in one apartment of "a sort of underworld country," men are preparing to eat: "And it seemed to [Aaron] that what they were going to eat was a man, a naked man. But his second self knew that what appeared to his eyes as a man was really a man's skin stuffed tight with prepared meat, as the skin of a Bologna sausage." This hollow/stuffed man is able to walk, and Aaron sees that he is "a big and handsome man in the prime of life, quite naked and perhaps stupid" (*AR* 286). Notwithstanding the fact that, in Richard Aldington's words, Lawrence "cannibalized the people and events around him" for his writing (qtd in Pecora 716), Lawrence, unlike Conrad, is not usually given to imagery of cannibalism. What are we to make of this bizarre trope? As Mark Wollaeger argues, cannibalism, "often suspected as Kurtz' 'unspeakable rites'" in *Heart of Darkness*, becomes in *The Secret Agent* "a more explicit metaphor for the horror concealed behind perceptions conditioned by convention" (142), a central modernist preoccupation. One of those screening perceptions is the integrity and apparent permanence of the self. The void is so frightening in Conrad because it symbolizes the nothingness that awaits all of us and that we, like Verloc on the verge of being murdered, refuse to see. Lawrence, acutely sensitive to the vulnerability of the self, could see the void all too clearly. Cannibalism in both authors (and, indeed, in a number of contemporary works)[33] suggests the disintegration of integral selfhood in a particularly graphic way, as explosions do in another. Anthropophagy implies a horrific and extinguishing merging with the devourer. In the modernists, the motif is not yet literalized as it will be in fiction after the Second World War; it remains metaphoric with varying degrees of explicitness.

In *Aaron's Rod*, the cannibalistic imagery is pervasive. On the surface, it is one more indication of a society gone rotten, but lurking beneath the textual rendition of cannibalism is the subtextual anxiety related to the devouring mother.[34] Lottie, in her resentment at the importance her husband places in the public world to the detriment of the domestic, reproaches Aaron for his concerns regarding the miners: "'you eat your heart out about it. More fool you'" (*AR* 8). At the performance of *Aïda*, the "fleshy" women in the opera "all looked such good meat" (*AR* 46). Sir William's aide, Arthur, looks "as if he would make a marvellous *bonne bouche* for a finely-discriminating cannibal" (*AR* 169). The narrator or Aaron sees the sterile sameness of modernity as "the big oneness swallowing up the many small diversities in its insatiable gnawing appetite" (*AR* 152). Despite this variety, however, it is clearly woman's love of man that is the supreme devourer in *Aaron's Rod*. The narrator comments on Jim's parodic eating of the body and blood of Christ, for instance, that the modern "holy Communion" is the sacramental "eating" of man by woman, after which "the blood and the body of which she has partaken become insipid or nauseous to her, she is driven mad by the endless meal of the marriage sacrament" (*AR* 165–6). In his crushing illness after giving way to Josephine, Aaron tells Lilly that he is "'nothing but a piece of carrion'" (*AR* 93). Likewise, he believes that the Marchesa will "drink the one drop of his innermost heart's blood, and he would be carrion" (*AR* 273). For Manfredi, love "'eats us inside'" (*AR* 242).[35]

In Conrad's *The Secret Agent*, the theme of disintegration, developed by cannibalism imagery among other techniques, is the keynote. All the major players of the novel disappear into nothingness. Even the Professor, who, while counting his own self to be of little worth in the struggle for the "regeneration of the world," paradoxically seems to possess one of the strongest personalities in the book, disappears at the end of the novel into the anonymity of a Joycean crowd: "Nobody looked at him. He passed on unsuspected and deadly, like a pest in the street full of men" (269). A virus or bacterium may be deadly, but it can hardly be said to have a personality or selfhood. Lawrence, certainly, was also capable of imagining the dissolution of selfhood, and his use of cannibalism as a motif in *Aaron's Rod* may very well be related to his feeling of vulnerability.

Hayles discusses *Women in Love*'s portrayal of confinement, as in the chapter "Snowed Up," as "the reduction of the enclosure" that opposes "expansion into the infinite." Such a confinement leads in the individual to "a cannibalistic feeding on the interior life of the self because one has failed to break out of the confines of the self" (102). In *Aaron's Rod*, there is clearly no desire to break out of the self's confines; quite the opposite, and for that reason, *Aaron's Rod* posits a more positive, if less convincing, outlook than either Conrad's work or *Women in Love* does.

The threats to selfhood represented by the "cannibalism" of modern society are balanced by the space imaged most significantly by the aesthetic space of statuary, which represents human (read male) centrality and selfhood. It is significant that, when Aaron arrives in Florence and encounters for the first time the Piazza della Signoria with its (copy of) Michelangelo's David and the sculptures of Bandinelli in "the dark, strong, inviolable square," which itself seems a metaphor for integral selfhood, he feels that he has arrived at "a perfect centre of the human world." And immediately, "Aaron felt a new self" (*AR* 212), buttressed as he is by the representation of selfhood in the statuary. Here, as Howe suggests, "the self is objectified in the body of man" (89). Several times in the novel, Aaron seems to aspire to the condition of statuary himself. When he watches the ritual dancing around the Christmas tree, Aaron stands "unmoving, static in everything" (*AR* 35). He seems as inert, mute, and spellbound as a statue. Just before he is about to embark on his affair with the Marchesa, Aaron remains standing at his window for a full hour: "Motionless, planted, he stood and watched that terrace across above the Arno. But like a statue" (*AR* 258). In something of a reversal of the Pygmalion myth, the ideal for humanity seems to be to aspire to the condition of statuary. The David seems to represent an ideal that Aaron at this point realizes he cannot attain: "'I can't stand by myself in the middle of the world and in the middle of people, and know I am quite by myself, and nowhere to go, and nothing to hold on to'" (*AR* 246).

Baker identifies the big, handsome sausage-man of Aaron's dream as Aaron himself, and that seems reasonable, particularly given the fact that the dream man is in the prime of life and is perhaps a little stupid. And we recall the Marchese's claim that love "eats us inside." The David statue is also big and handsome but emphatically not in the prime of life. David is a boy, an adolescent, as Lawrence makes clear in his essay "David" (first published in 1923), where he continually emphasizes the youthfulness of Michelangelo's figure (*P* 61–3).[36] Adolescence is a point in human development at which differentiation of self is about to culminate. Indeed, the essay calls David "[t]he pride of the fulfilled self … the full tree of life in blossom" (*P* 64). It is, perhaps, not so strange for David Herbert Lawrence to regard as an emblem of selfhood a statue bearing his own first name.

However, it is in the centred selfhood of aesthetic space that the splitting of the experiencing consciousness into Aaron and Lilly can be seen to be damaging to the novel. The quest for selfhood has been Aaron's, and in Florence he seems to have a chance, according to the novel's own standards, to achieve something important within himself. Lilly reminds Aaron towards the end that "'[y]ou are your own Tree of Life, roots and limbs and trunk'" (*AR* 296). But once Lilly reenters the scene, it is he who commands centre stage and thereby displaces Aaron, as the biblical Aaron was displaced by his younger brother, Moses. What Aaron sees in Lilly is "an intense and satisfying sense of centrality … alert enjoyment of being central, life-central in one's own little circumambient world" (*AR* 292). Unlike the David, however, Lilly has no aesthetic space around him that would allow for projective identification. He demands submission. Aaron may aspire to the same condition of self-centredness, but the message at the end of the novel is that he will never be Lilly's equal, at most an acolyte. In this way, the central theme of sufficient singleness in *Aaron's Rod* is seriously undermined by the "splitting" of the perceiving consciousness into the doubleness of Aaron and Lilly.

Aaron's Rod, perhaps more so than *The Lost Girl*, anticipates the leadership works by arguing that the philosophy that underpinned the marriage novels no longer accords with modern reality, if it ever did. The Arctic/African polarity of *Women in Love*, meant to describe the two great drifts towards racial/cultural demise in which dissolution is made inevitable by the break with the balance between body and mind (*WL* 253–4), was counterposed by the idea of star equilibrium in a heterosexual love relationship whereby a balance would be achieval between the need for singleness and the acceptance of connection. Now both give way to a new formulation enunciated most clearly by Rawdon Lilly. The two antithetical urges in humanity are now called love and power, a polarity that serves much more clearly than the earlier Arctic/African opposition to bring out Lawrence's fear of engulfment. Lilly makes it clear that the love urge (whether directed at an individual, at humanity as a whole, or at God) leads to a loss of selfhood, whereas the power urge, he tells Aaron, protects the self: "'You've got an innermost, integral unique self, and since it's the only thing you have got or ever will have, don't go trying to lose it … Your own single oneness is your destiny'" (*AR* 295).

In *Kangaroo*, the first novel of the leadership period *per se*, the way in which Lawrence's narrative strategy involves self-contradiction becomes obvious. In that novel, Ben Cooley initially represents the power urge, at least in his political platform and his tactics. But Lawrence simply does not know how to end the novel without the inconsistency of also associating Cooley with the love urge so that, when the Lawrence figure, Somers, rejects him, Somers is rejecting love rather than power. Cooley's character was based on an Australian Jew, Major-General Sir Charles Rosenthal (Darroch 61), and Lawrence simply made him Jewish in this hastily written novel. But since Jews for Lawrence are associated with the threat of merger and of boundary violation, Cooley illogically becomes an exponent of the love urge. In Lawrence's brilliant *Studies in Classic American Literature* (1924), we see further examples of the way love seems to threaten the self through merger with an other. In the essay on Poe, the fate of Roderick Usher is clarified through the perspective of love's dangers: "Love is the mysterious vital attraction which draws things together, closer, closer together," but there must be limits to the merging, for "[t]he central law of all organic life is that each organism is intrinsically isolate and single in itself" (*SCAL* 71). Lawrence even issues a warning later on in the essay: "Beware, oh woman, of the man who wants to *find out what you are*. And, oh men, beware a thousand times more of the woman who wants to *know* you or *get* you, what you are" (*SCAL* 76). Of course, the symbolic (or actual) brother-sister incest in Poe's story represents the dangers of love-merger and boundary violation: "They would love, they would merge, they would be as one thing. So they dragged each other down into death" (*SCAL* 85). And "the result is the dissolution of both souls, each losing itself in transgressing its own bounds" (*SCAL* 86). Similarly, when writing on Melville, Lawrence argues that "[e]very relationship should have its absolute limits, its absolute reserves, essential to the singleness of the soul in each person" (*SCAL* 152). As for Whitman, Lawrence's chapter on the poet, as one would expect, reads like one long tirade against love and merging: "This merging, *en masse*, One Identity, Myself monomania was a carry-over from the old Love idea. It was carrying the idea of Love to its logical physical conclusion" (*SCAL* 182).

Though Lawrence saw *Aaron's Rod* as the end of his "serious English novels" (*L* iv 92),[37] its ending is notoriously a non-ending, more like the middle of a conversation. Pinkney has compared *Aaron's Rod* to Fielding's *Joseph Andrews*: "An attractive, socially humble, would-be chaste hero who, wherever he goes, finds sexually voracious middle- or upper-class women eager to bed him" (112). To this, one may object that Joseph Andrews turns out to be of gentle birth, that is, to have a hidden origin in a privileged class. In Lawrence's novel, the mystery of origin has become the puzzle of ending. The author does not know where he is to go in his chosen form of art, but he knows where he has come from. The self-reflexivity in *Aaron's Rod* comes from Lawrence not quite knowing his direction and being in a void of sorts creatively; the aesthetic space in the novel is necessary for Lawrence's art to continue no matter what we may think of the results.

As William R. Barr argues, "[t]he narrator of *Aaron's Rod* … calls attention to the novel as fiction as much as to Aaron" (223). Gerald Doherty (who views *Aaron's Rod* as diachronic, sequential, and metonymic rather than spatial) claims that the novel, along with *Kangaroo* and *The Plumed Serpent*, "enact[s] a radical break-out from the close spatial confinement of the earlier texts" (*Theorizing* 25–6). The assertion suggests an interesting metafictional context for the motif of confinement. It is possible, in short, to see both the character Aaron and his rod as a synecdoche for the novel *Aaron's Rod*. The book's title and central totemic object are the same, just as Lawrence referred to his story *The Fox* as if it were a vulpine creature. As Barr argues (223), the destruction of the flute in *Aaron's Rod* also implies that Lawrence is sardonically applying to his own work an idea formalized in his essay "Surgery for the Novel – or a Bomb" (first published in 1923) to the effect that the novel has "got to break a way through, like a hole in the wall" (*P* 520), while the adaptations of the picaresque form suggest "an awareness of the need for and possibility of literary evolution" (Barr 223). In Dr Lilly's "surgery," the author attempts, so to speak, to massage the novel form back to life; in Florence, he explodes it; it will no longer do. As Lawrence had said earlier in a letter to Bertrand Russell, "One can still write bombs" (*L* ii 547).[38] As Robert Kiely argues, for Lawrence "when design became an end in itself, inflexible and absolute, it undermined the highest purpose in art" (93). One can bomb one's own writing when it becomes atrophied, but, unfortunately, one's writing can also bomb. In Aaron's "pillar-of-salt" returns to his home, Lawrence is revisiting his focus on marriage in *The Rainbow* and in *Women in Love*. The scenes suggest, as does the destruction of the flute, that Lawrence in *Aaron's Rod* is winding up old and now-painful material before moving on to he knows not quite where.[39] But the novel form he has been using has served its purpose by the end.[40] As Lawrence has Aaron think, "[t]he bomb had settled it" (*AR* 288). The ultimate unintended irony in the novel is that in clearing a space for his leadership works at the end of the transitional period, Lawrence embarks on a journey that will take him to a closed space, a dead end, in part because he will refuse to heed in his future work the present indications that the new ideas may not be as valuable as they appear. While the broken blue ball and the broken flute may suggest that the marriage phase is finished, when the windbag Lilly has the wind punched out of him and when the bomb at the café in Florence puts an end to Lilly's talk, we are seeing Lawrence's own doubts surface. As Vine suggests, in the café scene, "both Lilly's Lawrentian discourse and Aaron's rod are shattered" (xxxiii), and "Aaron's rod – like *Aaron's Rod* – ends with a breaking rather than a blossoming of the life that it seemed metaphorically to promise" (xxxiii-xxxiv). As we will see in the next chapter, a shorter fictional form arrives at the same conclusion.

4 Death and Survival in the Stories of *England, My England*

Like the three *Ladybird* novellas, many of the stories in the volume *England, My England* were written and even published earlier than their collective publication by Thomas Seltzer in 1922 would indicate. The title story of the volume, "England, My England," for instance, was first published in the *English Review* of October 1915, a few months after its composition. As was the case with *The Captain's Doll*, *The Fox*, and *The Ladybird*, Lawrence refashioned many of these stories a few years after the end of the war – in the case of "England, My England," the revisions were complete by December 1921 – turning them into stories that have much to tell us about Lawrence's preoccupations and narrative strategy in the transitional period.

In this chapter I will be concentrating on four stories from *England, My England*, three of which are war tales, specifically the title story, where Lawrence's use of evolutionary concepts is possibly as clear as in anything he wrote; "The Blind Man," about a soldier returned to his wife from the battlefront in a wounded or scarred condition much like Basil in *The Ladybird*; and "Hadrian" (also known as "You Touched Me"), where a rather Heathcliffean adopted son becomes a soldier during the war and returns home on leave (like Henry Grenfel of *The Fox*) to wreak havoc in his adopted home. I will also be presenting a reading of "The Horse-Dealer's Daughter," which, although not a war story in terms of its setting or overt concern, was written and revised in the same immediate postwar time as most of the other tales in the volume. It appears to go against the grain of the other fiction of the transitional period (except perhaps for *The Lost Girl*) in that a female figure is saved rather than destroyed or rejected. For that reason, among others, it is important to deal with "The Horse-Dealer's Daughter": it may be viewed as the exception that proves the rule, although, as I hope to show, this much-anthologized and well-known story is not as exceptional in this regard as it might at first appear to be. Before embarking on an analysis of these four stories, however, I would like to briefly examine selected aspects of three other tales in the collection.

In "Samson and Delilah," the Samson character, Willie Nankervis, might well be a portrait of Aaron Sisson returned home after many years (sixteen in Willie's case) in recoil from the roles of husband and father. Willie has been in America, and the sorry state of the Cornish mining town to which he returns in the first year of the war is meant to strongly suggest an economic motive for Willie's long-ago departure: "Tall, ruined power-houses of tin-mines loomed in the darkness from time to time, like remnants of some by-gone civilisation" (*EmyE* 108). However, Willie never offers an excuse for abandoning his family, despite being strongly condemned by his wife, Alice, who is backed up by a chorus of soldiers staying at her public house. Willie is the Lawrentian male who will assert his rights, stand his ground, and do whatever pleases him, conventional morality be damned. Where Aaron Sisson is buffeted by a series of women and lands flat on his back, Willie insists "he was on his own ground" even in his greetings to the miners who pass him in the street. He has "a sense of mastery and of power in conflict" (*EmyE* 108). But Willie does end up in a horizontal position when he is attacked by Mrs Nankervis, ably assisted by three soldiers, who tie him up and eventually take him outside (in the scene meant to echo the capture of Samson by the Philistines with Delilah's help). Willie is not only subdued, then, but, after sixteen years of presumed freedom, he returns to the kind of confinement that sent Aaron on his way and that we may be fairly confident propelled Willie as well. If the title of the story is anything to go by, Willie's future with his wife is not going to be a bright one, and his return may suggest an ironic blindness that looks both to Samson's fate in the bible and to the companion story "The Blind Man."

The war of the sexes in this tale, as in others of the period, reflects and parallels the larger conflict taking place in Europe. Like some of the soldier characters in the fiction of this period, Willie returns home wounded, but in this case it is an old collarbone injury from the mines that does nothing to take the fight out of him. The problem for Lawrence, however, may be that women, too, are prepared to fight; Willie is not the only one who savours "power in conflict." His wife decides to "gather her forces" (*EmyE* 114) like a general, and, indeed, does plan the attack and direct the three soldiers in the campaign to subdue Willie. At closing time, she announces "'Gentlemen – the enemy!'" (*EmyE* 112). Lawrence describes her as something of an explosive force ("something had exploded in her" [*EmyE* 113]), but Willie is "impenetrable" (*EmyE* 114). She even tells her husband to "'be a *man*, not a brute of a German'" (*EmyE* 117). In short, this is war. Trudi Tate has remarked that while many of Lawrence's contemporaries (Ford, Woolf, Rebecca West, for example) "tend to represent soldiers as traumatised or war neurotic, Lawrence's soldiers are often remarkably serene; indeed, many are less disturbed than his civilian characters" (102). The reason for that equanimity, I think, is that Lawrence's soldiers serve a symbolic evolutionary interest, and Lawrence is only very little concerned with their war experiences as experiences. When he tries to portray a battle realistically, as in "England, My England," it is not very convincing as war reportage. One ought to add, however, that the serenity identified by Tate as a characteristic of the Lawrentian soldier is often accompanied by a ruthlessness that is connected to the war in complex ways.

"Tickets Please" is usually taken as a narrative about the battle of the sexes that was the predictable result of the gender dislocation in Britain during the war. Written for the most part in a jaunty, superficially light-hearted manner, the story culminates in a serious attack on a young male tram inspector by a group of young female tram conductors. It shares some Darwinian features with the other war stories of the collection. Here, the war of the sexes rages on, and yet another male protagonist finds himself in an involuntary horizontal position ("At last he was down" [*EmyE* 43]). John Thomas Raynor, the tramway inspector, who, for some unspecified reason, is not off at war like other men his age, supervises a group of female tram conductors who clearly owe their jobs to the war and to the absence of men. The rhetoric suggests something masculine about them.[1] The women are described as "fearless young hussies" who "have all the *sang froid* of an old non-commissioned officer" and "a sailor's dash and recklessness" (*EmyE* 35, 36). "Tickets Please," then, raises questions about manhood and male-female relations in a time of turmoil, but the gender war works to some extent to mask an underlying survival anxiety.

John Thomas's overthrow from his position as a "fine Cock-of-the-walk" (*EmyE* 36) is organized by Annie Stone, one of his jilted conductors. She arranges it so that all of John Thomas's "old flames" (*EmyE* 40) among the tram conductors (and the fire metaphor is significant throughout the tale) are gathered in their waiting room ready to pounce when the inspector shows himself. He duly arrives, and they begin his torment with a game of touch. He is to stand facing the wall and guess which woman touches him on the back. Instead, Annie delivers a hard blow to his head, and the attack by the six women is on in earnest, at one point getting out of hand to the extent that John Thomas's physical survival seems to be in question. As in "The Blind Man," "Hadrian," and "The Horse-Dealer's Daughter," touch is important in this story, but selection or choice is the keynote.[2] The women decide that John Thomas must end his tom-cat ways and select one of them for a mate. The word "choose" or a variation of it is repeated a dozen times in the last few pages. The victory for the women turns out to be a Pyrrhic one, however, as they all, and especially Annie, feel little triumph at the end. In this sense, "Tickets Please" is able to maintain a remarkable balance between the sexes unusual for the period.[3]

Kiernan Ryan's materialist reading of "Tickets Please" ably brings into relief its subversive potential with respect to the institutions that support traditional gender roles: "Lawrence is intent on prising up and profiling the internalized sexual roles and conventions otherwise undiscerned because countersunk flush with the surface of everyday consciousness" (506).[4] The story "'denaturalizes' Annie's 'falling in love'" and shows her behaviour to be conditioned by internalized norms and the force of the attack on John Thomas to be a function of the extremity of the oppression of women of the time (508). This interpretation nicely accounts for the ambiguity of the ending. Ryan helpfully suggests that the touching game that precedes the actual attack and the insistence that John Thomas choose one woman of the six is a parody of the sexist conventions that hold the women in thrall: "The game holds up a mirror to the women's debased status as market commodities … John Thomas has transgressed the basic code of the sexual market by sampling and soiling all the wares then refusing to buy" (509). The revenge can never be satisfying, argues Ryan, because it springs out of the same dehumanizing market ethic that propels John Thomas on his career of exploitation. Compelling the inspector to choose backfires on the women, for it merely reinforces the male prerogative in the existing social structure (511).

Persuasive as Ryan's reading is in part, it can maintain its position only by ignoring to a large extent what Lawrence is doing with John Thomas himself and the extent to which the author identifies with his male character. Richard P. Wheeler's psychological reading makes a foray into those areas and in so doing touches upon the importance of survival to an understanding of the story. In Wheeler's view, John Thomas regains some of his dignity and power at the end. He emphasizes the role in the development of the story of Lawrence's experience as a young man at Haywood's surgical appliances factory when Lawrence was mobbed by a group of young women who attempted to pull his trousers off and expose his genitals: "The survival of phallic John Thomas seems to offer magical assurance that the object of the Haywood's girls' search has not only survived their attack but has been transmuted into an unconquerable folk deity" ("Cunning" 244). The one to be inspected has become the inspector, in other words. Wheeler probes the possible genesis of a character like John Thomas in the dangers inherent in a mother-child relationship that, like Lawrence's relation to his mother, is always threatening merger and thus the annihilation of the separate existence of the child: "Relations of intimacy pose a special threat, since the ideal of autonomy must be maintained by a personality endowed with weak or contingent self-boundaries" ("Cunning" 245). The tram conductors, having become "masculinized" women, threaten the manhood of their inspector, whose only recourse seems to be to lord it over the women phallically.[5] He must keep the relationships impersonal in order to protect his vulnerable ego boundaries. The open spaces of the fair at which Annie and John Thomas initiate their fateful relationship contrast with the enclosed (and potentially confining) female spaces of the tram and the waiting room (Wheeler, "Cunning" 245, 246). The attack itself "dramatizes the hostile, jealous side of the maternal identification" and its violence suggests "a boy's terrified helplessness before maternal hostility." The attack on John Thomas the man is also, of course, an attack on *the* "John Thomas," the penis, and is thus a symbolic castration: "But castration anxiety is subordinated to the deeper threat to the rigid but brittle boundaries of the narcissistic self" ("Cunning" 247). We will see some of the same anxiety to protect the threatened self at work in all four of the stories we will be looking at shortly but most graphically in "The Blind Man." Wheeler agrees with most commentators on "Tickets Please" that a remarkable degree of artistic control is evident: despite Lawrence's obvious personal investment in aspects of the story as he explores his own emotional conflicts, he creates a narrative at one remove from himself ("Cunning" 249).

I would argue that this story also explores in an unobtrusive way the sexual selection at work when an environment has changed. The key factor in the Darwinian conception of fitness is chance. Chance determines which traits will be "selected" as useful in any given environmental condition for the purposes of survival and mating. In the words of T.H. Huxley's *Evolution and Ethics*, the result of the struggle for existence "is the selection, that is to say, the survival of those forms which, on the whole, are best adapted to the conditions which at any period obtain" (Paradis and Williams 62). In this story, chance has determined that most of the young men in the town are away, and so it seems that John Thomas has an environment where he has few rivals and where the power of sexual selection is his to wield. It does not turn out quite that way, however, as chance also seems operative in the unplanned meeting at the Statutes fair of John Thomas and Annie, who turns out to be his favourite among the women and his nemesis: "Who should be the first to greet Miss Annie, on the show ground but John Thomas" (*EmyE* 37). By the end, John Thomas has been subjected to a "wounding" that links him to his absent comrades on the front and that restores a balance of sorts in the gender relations during the crisis time of the war: "His face was bleeding with a long scratch, his brow was bruised" (*EmyE* 43). The Darwinian sexual selection at the heart of the plot raises troubling doubts with regard to its fellow evolutionary mechanism, natural selection. John Thomas's fitness to live has been questioned – "'You ought to be *killed*, that's what you ought,' said Annie, tensely" (*EmyE* 44) – and there seems to be a powerful conflation of guilt about not going off to war like other men, doubts about sexual and gender identification, and fear that the fragile self will not survive as a viable entity. These are Lawrence's survival anxieties, and in "Tickets Please" he explores them through the evolutionary devices that allow him to treat his characters with some ruthlessness as well as some understanding. Lawrence, repeatedly *not* selected for military service but feeling under attack from all sides, may also have seen an aspect of himself as writer in the bloody but unbowed inspector who fights to retain the power of selection.

Keith Cushman has argued that *England, My England and Other Stories* is outstanding among Lawrence's published volumes of short stories and that there is a unity among the stories because of common themes, structures, and tone.[6] He suggests that one of the strengths of the book is a "sardonic detachment" that is lacking in other Lawrence stories (491). Insofar as "detachment" means a ruthless narrative strategy that applies Darwinian methods to plot and character, I would agree. Cushman's observation that Lawrence "seems almost to take delight in observing his characters as they march toward perdition" (492) would seem to support such a view.[7] The "tight, strong" structure that Cushman sees in so many of these stories is due not only to the use of myth and fairy tale in Lawrence's project of "revisionary mythmaking" as Cushman argues (493) but also, I would maintain, to the theme of survival that produces much of the energy in the stories. That very theme ensures that Lawrence is by no means as detached from the developments of the tales as Cushman suggests he is.

As these are war stories for the most part, the omnipresence of the threat of death is natural and does not strike the reader as anything out of the ordinary, but, in some cases, the presence of the war screens the underlying source of anxiety. "The Horse-Dealer's Daughter," which deals with dead parents and a woman who means to die (as we eventually learn), is not a war story as such but emerged out of the despair that enveloped Lawrence during and for some time after the war. The emphasis on death in the collection may be related to the war in many cases, but Lawrence's survival anxiety is a powerful force that preceded the war and was simply exacerbated by its onset. That this is so becomes clear if we look at "The Primrose Path," which was written before the war and only slightly revised for the 1922 publication. In a harrowing death-bed scene, Daniel Sutton, a character based on Lawrence's maternal uncle Herbert Beardsall (*EmyE* 246n), terrified at witnessing illness and death, visits the wife he abandoned many years earlier and stands over her "like one condemned" (*EmyE* 130). Sutton's wife is dying of consumption, but this is only one of a series of displacements that allows Lawrence both to acknowledge his own fear and to distance himself from it at the same time. Cancer is mentioned in the story, but it is the mother of Daniel Berry, Sutton's nephew, who has died of that disease, news of which intensifies the uncle's death anxieties. Berry is a Lawrence stand-in of sorts and bears much the same relation to Sutton as Lawrence did to his uncle Herbert. The dying woman is clearly also Lawrence's own cancer-stricken mother on her deathbed. There is even mention of Sutton's belief that the woman with whom he had eloped to Australia tried to poison him, a possible projection of Lawrence's mercy killing of his mother.[8] The horror of death has simply been transferred from Daniel Berry to Daniel Sutton, who not only has the same first name as his nephew but even makes Berry feel "as if he were sitting beside an older development of himself" (*EmyE* 125). Berry's thoughts and emotions are largely effaced, and he is content mostly to observe his *doppelgänger* deal with life's vicissitudes, thereby screening him. As Sutton stands over his wife, he becomes her young husband again for a moment and even, significantly, becomes once more "a child in horror of that protruding mouth" (*EmyE* 130), thus suggesting that a filial relationship lies behind the horror of the scene.

As the self-effacing observer, Berry is a Lawrence figure in the metafictional sense as well. He views his uncle from an odd, emotionally cold distance, almost as if the older man were a specimen of life (or a fictional character) to be studied. As Weldon Thornton suggests, Daniel Berry "enjoys putting people into emotionally awkward situations and observing their response" (20), a description that might describe the ruthlessness that Lawrence the author develops into a fine art in the transitional fiction. We may take the ending of the story to be a reflection of Berry's detachment, as Thornton does (20), but Berry's prediction of what will happen to Sutton in the future – "'That girl will leave him … then she'll go off with somebody else'" (*EmyE* 136) – may also suggest the author plotting his story out loud.

Even in the war stories of *England, My England,* I would argue, the emphasis is not so much upon the destruction of war itself as upon the pervasiveness of illness or injury of one kind or another in a society that blithely sends its young men off to be slaughtered. In "England, My England," Egbert seems to be diseased and degenerate even before he goes off soldiering. The war becomes a mere convenience for his drift to death. In "The Blind Man," Bertie Reid's aversion to physical contact is called an "incurable weakness" (*EmyE* 58) as if it were a disease rather than an unease. In "Hadrian," an old man's final illness is the trigger for a pervasive survival anxiety that propels the tale to its unexpected conclusion. "The Horse-Dealer's Daughter" is ostensibly about the awakening of a woman who, in a somnambulist state, walks into the water, but it is more subtly about a doctor who normally tends to other people's illnesses and now, having saved the woman from drowning, worries about his own state of health and the wet clothes that threaten his survival. As we will see, the difficulty of survival in a Darwinian universe is the central premise of the title story.

"ENGLAND, MY ENGLAND": THE FAILED VITALIST HERO

"England, My England" is the story of the dilettantish Egbert and his wife, Winifred, who, before the war, form a winsome couple living on little more than love. The birth of three girls seems to make the family complete, but underlying stresses, based on Egbert's disinclination to work and the interfering officiousness of Winifred's father, gradually sour the relationship. A crisis occurs when Egbert carelessly leaves a sickle lying about in the grass and Joyce, the eldest daughter, cuts her knee on it so badly that there is a danger she may lose her leg. When war breaks out, Egbert decides to enlist even though he has doubts about the legitimacy of the British war effort, and Lawrence makes clear that there is a strong suicidal element in Egbert's resolve.

The ending with Egbert killed in battle seems the inevitable climax to the story's direction from the beginning. In a sense, Egbert has been "killed" by Winifred's withdrawal of love, but beneath the tragedy of his death an extensive network of imagery points us to the conclusion that Egbert's death has also been the result of his failure to pass a Lawrentian survival test.

We can discern a number of similarities between "England, My England" and *Aaron's Rod*, likenesses that strongly suggest that the short story emerges out of the same general concerns that preoccupied Lawrence at this time in his life. There is, first of all, a central metaphorical contrast between a vertical strength and a horizontal weakness that is not found in the 1915 version. As in *Aaron's Rod*, the contrast derives ultimately from botany. A good, sturdy man who can stand on his own two feet is like a firmly rooted tree or plant that has a much better chance of survival than a poorly rooted one, which is prone to fall over the way Aaron does in his disease-reactive collapse after his encounter with Josephine Ford. In "England, My England," the horizontal versus vertical metaphorical system relates to the contrast between the weakness and passivity of Egbert and the strength and active nature of his wife's family, the Marshalls, in particular of the *paterfamilias* and Egbert's father-in-law, Godfrey Marshall. In a passage rendered through free indirect discourse, Winifred complains that Egbert "did not stand firm in the landscape of her life like a tower of strength, like a great pillar of significance ... He was like a flower in the garden, trembling in the wind of life, and then gone, leaving nothing to show" (*EmyE* 12). Winifred's mother even makes the connection to a specific flower, the lily, in a passage that features the same biblical allusion Lawrence uses in *Aaron's Rod* as the source of Rawdon Lilly's surname: "'consider the lilies, that toil not ...'" (*EmyE* 12). In general, Egbert is associated with such horizontal images as the garden path and, particularly, the sickle that "cuts down" his favourite daughter, Joyce, thereby rendering her horizontal, as it were. She is certainly unable to plant both of her feet firmly on the earth, for she is lamed. While the Marshall family is connected to vertical emblems such as trees and pillars, Egbert, almost literally, "stood for nothing" (*EmyE* 13).

We also see in the 1922 version but not in the original "England, My England" a strange description of sexuality in terms of lightning that also occurs in *Aaron's Rod*. Oddly, though, what in the novel is a description of male sexuality seems in the story to be Winifred's experience, a first indication, perhaps, that Lawrence wants to keep Egbert at a distance: "It was to her as if pure lightning, flash after flash, went through every fibre of her, till extinction came" (*EmyE* 13–14).[9] We might well view the association of lightning and female sexuality in the short story, a deviation in terms of Lawrence's normal gender associations, as the first major clue in "England, My England" that something is out of kilter in the marriage between Egbert and Winifred. Egbert is the kind of character that in Lawrence tends to be female: sensitive, refined, and hyperconscious. In retrospect, we might even venture to suggest that Egbert's failure to be a Lawrentian male of lightning strikes leads to an extinction far less pleasurable and much more literal than the one achieved by Winifred here. The "female" male (or a "male" female such as the woman who rode away or the protagonist of "The Princess") in this phase of Lawrence's writing, much in contrast to his earlier work, cannot survive in a harsh world of violence and extremes. Winifred does eventually become the typical Lawrentian female caught between the demands of body and spirit (manifested as duty) where one reality annuls the other, as we have seen many times before in Lawrence. However, this conflict within Winifred arises only after Egbert has reluctantly become a soldier. Both marriage partners are in false Lawrentian positions.

What turns out to be somewhat surprising is who does and who does not survive in the story. The protagonist functions as an antihero who clarifies Lawrence's other dynamic, vital dark men such as Henry Grenfel, Count Dionys, Rawdon Lilly, Ciccio, etc. Egbert is the failed Lawrentian vitalist hero whose drive to *thanatos* the ending seems to dramatize. However, Egbert is a very complex evolutionary failure for Lawrence. Like the David statue in *Aaron's Rod*, he is a Lawrence namesake, in this case bearing within his own name the appellation "Bert" from the author's Eastwood days. In many ways, his character is congenial to Lawrence, particularly when it comes to the competition between husband and child for the wife/mother, a recurring motif in the period that derives ultimately from Frieda's abiding sense of loss over her children.[10] And yet it is this Lawrence stand-in who perishes at the climax of the story. For, no matter how sympathetic Lawrence may be to Egbert in certain respects, the protagonist cannot survive, given the environment of the war, nor can his way of old Englishness outlive the conflagration. In effect, Lawrence is admitting that part of himself must die, has, in fact, been killed by the war and what the war represents. The rest may endure, but it will of necessity be a tougher, more ruthless Lawrence who survives as writer and man, one who is capable of penning "The Woman Who Rode Away," *The Plumed Serpent*, and other remorseless fiction.

Kim A. Herzinger makes a credible case for the concept that "'England, My England' is Lawrence's Parthian shot at both the Georgians and the England for which the Georgians stood. It is the fable of a failed Georgian; more precisely, perhaps, it is a fable about why Georgianism failed" (158).[11] The same Georgian attitude to the war embodied in Rupert Brooke is evident, according to Herzinger, in Lawrence's Egbert, who is "a prototypical Georgian who refuses the responsibility for his own life" (162). When thrown back on his own resources in a family crisis or in the larger crisis of the war, he finds that he has none (Herzinger 163). In contrast to the Georgian Egbert is the Victorian Godfrey Marshall, who, although old-fashioned and out of date, is palpably made of sturdier stuff (Herzinger 166). There is little doubt that Herzinger's reading of "England, My England" is essentially correct as far as it goes, but it also simplifies a complex text and in particular a highly ambivalent attitude to the protagonist evident throughout the story, an ambivalence that is the very key to the tale, as a reading from the point of view of evolution and survival suggests.

The largely subliminal themes of evolution and survival in "England, My England" are easy enough to spot once we are alerted to their existence. The story begins with Egbert working in and on his garden. He is denoted by nothing more than a pronoun for the first few paragraphs, a designation that foreshadows the ending, where he is but a dead face to the German soldiers. The first sentence reads: "He was working on the edge of the common, beyond the small brook that ran in the dip at the bottom of the garden, carrying the garden path in continuation from the plank bridge on to the common" (*EmyE* 5). The first time we see him, then, Egbert is attempting to extend human control (the garden) over nature (the common), and doing a rather bad job of it: he cannot get the path straight. His failure here is a synecdoche for his failure in life. T.H. Huxley's horticultural imagery explaining the human modification of the force of natural selection is a helpful gloss: "In the modern world, the gardening of men by themselves is practically restricted to the performance, not of selection, but of that other function of the gardener, the creation of conditions more favourable than those of the state of nature" (Paradis and Williams 101). Egbert fails signally to create conditions more favourable for himself than "the state of nature" because such a refined man, in Lawrence's view, has little chance in barbarous times.[12] This is true in both the 1915 and 1922 versions. In the later version of the story, significant implications stemming from such gardening scenes combine with the botanical metaphorical system discussed above. The result is that this opening scene suggests in terms of evolutionary influence an analogy both to the Darwinian breeder at his eugenic task and to the author tending to his fiction. At least on one level, Bert Lawrence is attempting to connect the garden path of his past fiction to his present savage heath of creativity.

The garden is not merely a controlled patch of growing nature; in this story, it is *over*refined, like Egbert himself, who is described as "purebred." It has been a garden for a thousand years. The common, correspondingly, is not only untended nature, it is described by words like "savage" and associated with snakes. We might say that Egbert is the personification of his beloved garden, and his failure to extend the garden into the wild common suggests the impossibility of his continued existence into the period of the savage war. He is unfit for modern life and battle. He cannot adapt, does not wish to adapt, just as he will not modernize the old cottage in the garden that he and Winifred come to inhabit. By the end of the story, Egbert is "weeded out" of the scheme of existence because of his unfitness. The carelessly discarded sickle that maims his daughter is an instrument of human control over unruly nature, but, ironically, it is Egbert who is cut down in the end on the field of battle. In the 1915 version, significantly, there is no blame attached to the child's accident.

In contrast to her husband, Winifred is robust and fit for survival. She is "ruddy" and "strong" and very much a Marshall. Her father is "a man of energy" with a "tough old barbarian fighting spirit" (*EmyE* 7) that would undoubtedly have seen him through the war had he been a younger man expected to enlist. His very *blood* is strong, and this makes for a "vigorous" home (*EmyE* 7). While the Marshalls' culture is "grafted on" (a telling horticultural image), the family seems, nevertheless, to have achieved a rough balance between spirituality and materiality: "If there was poetry after dinner, there were also chocolates, and nuts, and good little out-of-the-way things to be munching" (*EmyE* 7). They are northern people, while the southerner Egbert is "made of quite a different paste. The girls and the father were strong-limbed, thick-blooded people, true English, as holly-trees and hawthorn are English," while Egbert is "a born rose" (*EmyE* 7) whose culture has been internalized over many generations of breeding.

Lawrence's use of trees, buds, seeds, et cetera as metaphors is so prominent throughout his writing career that the tree must be accounted his primary model of life. In this transitional phase of his canon, however, the model is given an evolutionary twist. In *Aaron's Rod*, the biblical background to the rod has botanical references, and the tree symbolizes the ideal of male rootedness. In *The Fox*, the tree becomes both an image of the sterility of the March-Banford relationship and the instrument of a phallic substitution for that relationship. In "England, My England," the tree and related botanical metaphors are the primary means by which Lawrence's Darwinian interests are discernible. Winifred's "family tree was a robust vegetation that had to be stirring and believing." Consequently, Godfrey Marshall, who is made of "tough, rough fibre," has none of the misgivings about society and religion that make his son-in-law so diffident. His faith is "an acrid faith like the sap of some not-to-be-exterminated tree ... Perhaps he was unscrupulous, but only as a striving tree is unscrupulous, pushing its single way in a jungle of others" (*EmyE* 14–15). Lawrence means the contrast between Godfrey and Egbert in terms of survivability to be stark. But at this point in the description, while continuing and extending the botanical references, Lawrence embarks on one of the confused and confusing passages that characterize his writing of this period. "In the end," maintains the narrator,

> it is only this robust, sap-like faith which keeps man going. He may live on for many generations ..., as pear-trees and currant bushes would go on bearing fruit for many seasons, inside a walled garden, even if the race of man were suddenly exterminated. But bit by bit the wall-fruit-trees would gradually pull down the very walls that sustained them. Bit by bit every establishment collapses, unless it is renewed or restored by living hands, all the while. (*EmyE* 15)

This renewing is what Egbert fails notably to do. So much is clear, but the metastasis of metaphorical references in the paragraph is so marked that we start to wonder if we are not barking up the wrong tree. At first, "man" apparently *is* the tree, but then he is the gardener tending the tree. We begin with sap (rather than the blood one might expect from Lawrence) as the image for the faith that keeps one going, the way sap, presumably, keeps the tree growing. One may survive physically, protected behind the "shelter of the social establishment" for a time (*EmyE* 15), just as fruit trees may go on producing behind garden walls even if the gardener (whose pruning really gets the sap going in the trees and bushes) disappears along with all his race. But instead of the metaphor continuing with the trees and bushes eventually succumbing without such tending, it is the protective walls that crumble (taking down the espalier trees with them, admittedly). And in the midst of an extended metaphor that is meant to explain how, above and beyond the "social establishment," faith is needed for humanity to thrive, we have an image of the extermination of the entire human race. Further, we are told that plants need human hands to thrive, but humans need sap. What is more or less clear is the implicit claim that the individual needs the pruning of adversity to keep the sap of life running nicely, and that the gardener, a variant of the Lawrentian gamekeeper figure, is a metaphor for the novelist tending his fictional garden, eliminating the deadwood that fails to survive. In the garden, there will always be those trees able to push their way "in a jungle of others" and thus survive. And there will be those trees that cannot do so and will not survive.

Godfrey Marshall is an extreme example that clarifies a general tendency in the transitional period with respect to the portrayal of more than one generation. Perhaps the only other older-generation character with a comparable role in the period is Mr Houghton of *The Lost Girl*. In the three novellas, there are Henry's grandfather and Banford's father in *The Fox* and Lord and Lady Beveridge in *The Ladybird*, all relatively or very minor characters. Captain Hepburn's wife is a pseudo-mother, as we have seen. It is not often that we have three generations depicted as completely as we do in "England, My England." The presence of more than one generation, in any case, allows Lawrence to deal with the issue of "blood" and inherited traits, desirable and undesirable, a central Darwinian concern. The obvious contrast to this tendency in the transitional period is what Lawrence does with the generations in *The Rainbow.* There he seems interested in how certain "non-human" traits – the "carbon" – tend to repeat themselves with variations in such a way that has consequences for self-fulfillment in a changing world. The interest in *The Rainbow* may be part of a continuum that leads eventually to the evolutionary jungle of *The Plumed Serpent*, but the emphasis in the earlier novel is still on identity and fulfillment, as it is in the 1915 version of "England, My England." Even though the revised story is in many respects identical or very similar to the original version, a crucial difference, *the* crucial difference for our purposes here, is that in the 1922 "England, My England" the focus has become survival, adaptation, and selection, concerns that are barely hinted at earlier. In this story, the focus is on a Darwinian analysis of fitness and adaptation based on traits bred in or out of a family.

The cottage in "England, My England" is called a place of breeding and mating. Godfrey Marshall, almost the Darwinian breeder figure in this respect, delights in the romance between his daughter and son-in-law in the early years of the marriage. When a child is born to them, the interest of the Marshall family might be characterized as eugenic in nature. Joyce is described in a way that suggests the importance for the Marshalls of the change that Egbert effects in what we would call the family's gene pool. Genetically speaking, in fact, Joyce is very much her father's daughter: "It was the first exquisite blonde thing that had come into the family … They had everyone of them strong, heavy limbs and darkish skins, and they were short in stature. And now they had for one of their own this light little cowslip child" (*EmyE* 10–11). There is nothing quite like this in *The Rainbow.* That it is Joyce who is maimed suggests perhaps that the father's unfitness has been passed on to the daughter.[13] In fact, she may well have died without the intervention of the Marshalls. That they too are prepared to act rather than let nature "take its course" suggests, in spite of some indications to the contrary, that this is far from a simple story in which Egbert represents culture and the Marshalls nature. But it is unmistakably a story of survival of the fittest.

In fact, in one key symbolic scene whose overtones resonate throughout the story, Winifred acts as the gamekeeper, as it were, and interferes in the order of natural selection. One day, she hears an unearthly cry and finds a snake in the process of gobbling up a frog whose one hind leg is all that remains visible. Winifred cries out, and the snake sullenly gives up its prize. One echoing image here is the leg of the frog, which looks forward to Joyce's injured limb.[14] It is also clear that the "normal" hierarchy of vitalism has been inverted in this savage time. The frog is a more advanced evolutionary being than the snake, but it is the lesser life that swallows up the greater in this case because the world is askew.[15] Winifred interferes in the natural order so that the victim or prey escapes death. The Marshalls do the same for Joyce, but Lawrence is not about to be so kind to Egbert. Although he greatly fears that, in a savage wartime world, he too is at risk, he must distance himself from his vulnerable protagonist in order to prove that he himself can survive.

The simultaneous textual identification with and rejection of Egbert creates an ambivalent text while Lawrence's anti-war feelings confuse the otherwise clearly demarcated evolutionary contrast between Egbert and Godfrey Marshall. Egbert's initial stand against the war certainly has narratorial approval and is consonant with much other anti-war prose, including the "Nightmare" chapter of *Kangaroo.* Even Egbert's lack of robustness – his floral fragility that is opposed elsewhere to the Marshall arboreal strength – takes on positive connotations in connection with the war: "when he was truly himself he could no more have been aggressive on the score of his Englishness than a rose can be aggressive on the score of its rosiness" (*EmyE* 27). In spite of his approval of Egbert's plans to enlist, Marshall too, like Lord Beveridge of *The Ladybird*, is constitutionally opposed to the jingoism and mob feeling whipped up by the war frenzy (*EmyE* 28). As a soldier, Egbert seems particularly ill-suited.[16] Even his military clothes make him seem out of place. His enlistment is portrayed as a self-betrayal and a suicidal gesture. Being a soldier even seems to make Egbert susceptible to illness: no longer "purebred" in appearance, he now looks "impure," develops a sore on his lip, and has "unclean" blood (*EmyE* 29). The man is ripe for Lawrentian extermination.

One of Egbert's problems in terms of the will to survive is that he has an allowance of £150 per annum that keeps him "from starving" but helps to drain all ambition from him. There is not in "England, My England" the same kind of emphasis on food and eating as in *The Captain's Doll*; a much different sort of survival is at issue here, but the obvious contrast to Egbert's subsistence patrimony is the state of Hannele and Mitchka in the novella, who, it is made clear, are forced to engage in survival strategies they would normally never consider because of the precarious, hand-to-mouth existence they lead in postwar Germany. They seem to be in real danger of starvation, while Egbert is like a domesticated predator made soft by the provision of food in a tin. Because Egbert has never before had to fight to survive, he fails to develop the necessary skills to do so when push comes to combat. Here, Lawrence, somewhat like Conrad in *The Secret Agent*, I think, is suggesting (in keeping with the premise of Max Nordau's influential book *Degeneration*) the general unfitness of the upper-middle class of English society, a class made weak by too-easy access to a bit of money but not free enough to pursue personal fulfillment. The 1915 Evelyn has the same income and the same diffidence as the 1922 Egbert, but the emphasis in 1922 with regard to this one issue of income is, suggestively, on the same kind of survival at stake in *The Captain's Doll*: "that fatal three pounds a week would keep him from starving as long as he lived" (*EmyE* 10). There is no necessity for him "to work for his bread and butter" (*EmyE* 14). In fact, if Egbert resembles anyone from *The Captain's Doll* it is the lamented Mrs Hepburn, at least the Mrs Hepburn that the captain portrays to Hannele as hopelessly out of her element in the modern world. That Egbert wishes to live in the spirit of old England and has no desire to "fight his way in the world" is meant to strongly suggest his unfitness for survival in a world where one must literally fight one's way.

Egbert's lack of "manly" aggression is also the nub of his marital problems, and the solution clearly looks forward to the leadership works. The narrator tells us that Winifred will "hunger all her life" for male power, having once tasted it in her father's presence, and that Egbert's abnegation of power makes her feel "like a fish out of water" (*EmyE* 17). The metaphors of hunger and suffocation suggest that her survival too is at stake, but it is not her fate that we follow. An interesting comparison with *The Fox* might be made here. Henry Grenfel, as we saw in chapter 2, denies or wishes to deny Nellie March any obligation and power of responsibility. Does she feel responsible for Banford? Well then, we simply get rid of the Banford. In fact, there is a marked emphasis in *The Fox* on the battle for responsibility. In some ways, it is the central conflict. In "England, My England," as in other matters, we get something of an inversion that begins to clarify why it is Egbert who does not and (ideologically speaking) must not survive. Particularly when it comes to child-rearing, Egbert forces Winifred to assume all responsibility, and yet he undermines her authority at the same time in the name of liberty. Again, the imagery is telltale and alarming, for the struggle between Egbert and Winifred over the children and the efforts of Winifred and the nurse to keep the children safe at the cottage are a virtual war: "From mother and nurse it was a guerilla gunfire of commands, and blithe, quicksilver disobedience from the three blonde, never-still little girls" (*EmyE* 17–18). The displaced warfare of family life will have to do until the real thing comes along; Lawrence, no doubt, would argue that the one actually leads to the other. But there is heavy irony in Egbert's refusal of responsibility when his carelessness leads to Joyce's accident. The description of Egbert "fatally" neutralizing Winifred's authority with the children proves almost literally true in Joyce's case. Winifred's father, as one would expect, takes charge in the crisis and is "implacable in his responsibility" (*EmyE* 23). The issue for the Marshalls as far as Joyce is concerned is that she be allowed to retain her "liberty of movement" (*EmyE* 26), an ironic phrase in light of Egbert's stand for liberty in the bringing up of the children.

By the time it comes, Egbert's death has been overdetermined by Lawrence's grudge against women and his fear of physical debility. The former is evident when, after Joyce is wounded, Winifred takes on the role of the "Mater Dolorata." Winifred feels her heart "killed and going cold" during Joyce's painful treatments at a London hospital. She is "like the Mother with the seven swords in her breast" (*EmyE* 23), like *The Ladybird*'s Lady Beveridge, another Mother Dolorata. It appears that Egbert pushes his wife into the mother role, unlike Captain Hepburn, who resists his wife's maternal control. The male who resists maternal power in Lawrence's scheme survives, while the male who succumbs to it or encourages it is doomed. Winifred also withdraws from her husband sexually, and this kills him as much as anything else. As for Lawrence's fear of debility, it is pervasively present in the wound trope. Even before Egbert's wounding and death in battle, all the major players in the family are wounded literally or metaphorically: Joyce in her knee, Winifred in her heart, and Egbert in his brow: "there was a little frown between his brow as if he had been cleft there with a hatchet: cleft right in, for ever, and that was the stigma." As a result of the change in family life, Egbert feels that the "horrible marsh-poison went through his veins and killed him" (*EmyE* 24). The metaphorical wound in the brow, of course, is linked to the head injury in battle that kills Egbert in actuality. We might say that Winifred and his own carelessness have killed Egbert as surely as Henry kills Banford in *The Fox*. Egbert's death in "England, My England" is not about war itself so much as the wartime conditions in England that determine survivability. The effete artist Egbert, so beloved of his wife early in the marriage, is doomed by an inhospitable England and a hostile wife. Lawrence is anxious to avoid a similar fate.

The 1915 and 1922 versions of "England, My England" are quite similar in terms of the general lines of the story and the major events: the marriage between Egbert (or Evelyn in the early version) and Winifred, the three daughters, the maiming of Joyce through the accident with the sickle, Egbert's enlistment in the artillery, and his death on the front. Lawrence's revisions make the later version a much better story, and, what is more, they clearly indicate a new-found interest in evolutionary concepts that seep into the narrative strategy. Godfrey Marshall's role is greatly expanded, to the point where he takes on the symbolic role of the robust, fit, strongly rooted tree destined to survive all adversity in contrast to Egbert's delicate, diffident, fragilely rooted flower fated to die on the field of battle. Egbert dies in both variants of the story, but in the 1922 version the death becomes the culmination of an exploration of evolutionary concepts that are both the result of the nightmare experience of the deadly war and the means, Lawrence believes, for him to get beyond a creative dead-end in his writing.

"THE BLIND MAN": MOTHER AND CHILD

The protagonist of "The Blind Man," Maurice Pervin, is more fortunate than Egbert of "England, My England." He has been to the front at Flanders but has returned to his wife, Isabel, albeit scarred and completely blind. The two are able to achieve a generally satisfying relationship that is, however, marred by occasional bouts of depression on his side and a maddening feeling of weariness on hers. When Isabel's old friend, the rather prissy bachelor Bertie Reid, comes to pay an extended visit, the unacknowledged problems come to a head, and emotional survival seems to be at stake, at least implicitly, as physical survival had been at stake for Maurice at the front. Maurice and Bertie dislike each other, as they must given Lawrence's narrative strategy of presenting them as embodiments of opposing attitudes to the outer world, and Maurice feels isolated during the after-dinner chitchat to the point where he retreats to the stable and to his animals for the feeling of warmth and security they provide. Isabel sends Bertie out to check on Maurice, and the story ends in something of an epiphany. Maurice touches Bertie's face and head and asks Bertie to touch his scarred face in return. The result is a feeling of exhilaration for Maurice and one of collapse for Bertie, whose own inadequacies have been described in terms of an "incurable weakness, which made him unable ever to enter into close contact of any sort" (*EmyE* 58), especially with women. This apparently simple tale of physical and spiritual strength triumphing over intellectual prowess masks a complex ambiguity for, as Tate has observed, "[a]n acute anxiety occurs at the very level of writing in this text" (104).

Analyses of "The Blind Man" have generally seen it in one of two ways. Kingsley Widmer's *The Art of Perversity* is characteristic of the view of the story as a fairly simple and typical Lawrentian "moral fable" that inverts the prevailing values of his society (44). The inversion is dramatized most obviously in the use of dark and light terms of reference and what they represent: "Lawrence's insistence on chaos, passion, withdrawal, the organic, and the acts of emotional violence, stand against self-consciousness, rationalized thought, social activity, and 'normative' behavior" (Widmer 46). What such a perspective fails to account for in any meaningful way, unfortunately, is the dramatically ironic ending, which we now know Lawrence added in revision for the initial July 1920 publication in the *English Review.* Mark Kinkead-Weekes tells us that the original manuscript of the story was completed by November 1918 and ended much differently than the published text.[17] The irony is "in a different direction" in the manuscript, and Bertie is handled much more gently. In fact, by the end, the two men feel "a new sense of being, a new kind of knowing and the blood-brotherhood, the total commitment to lasting relation, that Lawrence had always wanted with another man." As for Bertie individually, he "is surprised into a kind of knowledge, and commitment he had not been capable of before" (Kinkead-Weekes 480).

We must, I think, stand back from the story and attempt to examine Lawrence's narrative strategy in changing the ending so fundamentally. As with many of the other revisions of this period, the changes that Lawrence made in attempting to get his fiction to evolve into the desired new form make the text more compelling as well as more puzzling. The key question here becomes why does Lawrence choose to use biting, almost mocking, irony at the end, given what appears to be his intention in the story as articulated by Widmer and others?

Maurice's belief that he and Bertie have become friends is possible only because of his blindness, the very disability that the story otherwise argues is truly a strength in this inverted Lawrentian world. The other two characters know that Maurice's conviction that "'we've become friends'" (*EmyE* 63) is almost childishly mistaken: Bertie because he knows what his own feelings are and Isabel simply because she is sighted and is able to see at a glance that Bertie is severely discomfited. So Maurice's ingenuousness undercuts the obvious intention of the text to demonstrate the strength of "blood knowledge." Not only does the text seem to relish Bertie's collapse, it also seems to ridicule the male protagonist. How can we view the blind man's blindness as anything but a handicap in human relations, given his total misunderstanding of Bertie's reaction?[18] What is surprising too, when one views the story from the point of view of winners and losers, is that, although the struggle seems to be between the two men, with Isabel's allegiance at stake, Isabel herself has the most to win or lose. According to the light-versus-dark scheme, Isabel must be accounted a major loser in that her bid for a balance between light and dark, the "twilight" state invoked in *Twilight in Italy*, is frustrated. But from the perspective of the more subtle evolutionary scheme at work, she is the only winner. For that reason, Isabel is the key character in the story.

The second major school of thought on "The Blind Man" accounts for its ironic ending but fails, finally, to give Isabel her due. This alternative reading of "The Blind Man" is basically a Freudian one, certainly a psychological one, and it is ably represented by Richard P. Wheeler's analysis. Wheeler's premise seems self-contradictory at first blush: "Lawrence crushes the alternative represented by Bertie in order to protect the fantasy of complete intimacy with another that lies behind Maurice's experience in the story" ("Intimacy" 236). This is sound reasoning if we restrict Maurice's experience to his marriage, but the whole drive of the tale propels it in the direction of "complete intimacy" between the two men. That being the case, it makes little sense to crush the very other with whom that intimacy is to be created, or does it? Part of Wheeler's argument, to be fair, is precisely that Lawrence was unable to achieve what he longed for in terms of male bonding until *The Plumed Serpent*, so one might argue that the contradiction may inhere in the fiction as much as in the analysis of the fiction.[19] Wheeler suggests that both Maurice and Bertie unconsciously view Isabel as a mother figure, but that Bertie's relation to her is an idealized one that protects against the incest taboo while Maurice's "suggests a resolution of oedipal tension by a retreat to an earlier phase of psychic organization" ("Intimacy" 239,

240), more precisely, to that of an infant. Even though Isabel is the common denominator here, however, Wheeler seems uninterested for the most part in pursuing her perspective on things, perhaps because the Freudian approach tends to be so male-oriented. The relationship between the men is conditioned by a dual oedipal configuration where "each serves as symbolic father to the other" ("Intimacy" 241). In the Freudian reading, Maurice's blindness can be nothing but a symbolic castration even though the story makes clear that he is sexually attractive and potent and is several times compared to Freudian phallic symbols such as a tower. Certainly, it is true, as Wheeler points out, that Bertie is "an older man with a prior, exclusive claim on Isabel" ("Intimacy" 243), a father figure, that is, but this view avoids the question of why Isabel manœuvres the two men into the confrontation scene with each other, the result of which is the annihilation of her friend and the foolish exhilaration of her husband. When Maurice knocks off Bertie's hat in the barn in a preliminary approach to feeling his head, the Freudian equation of headgear and male genitals fits Wheeler's thesis of a father-son/son-father confrontation between the two men.

Because Freudian readings are very good with ambivalence (in that they recognize the power of projection and inversion), Wheeler's analysis of the story, unlike Widmer's, nicely accounts for the ironic ending. In Maurice's ironically loving destruction of Bertie we are dealing with a simultaneous longing for and fear of merger with another (Wheeler, "Intimacy" 246). The problem, though, is that although Wheeler does note that this ambivalence is rooted in infantile experience, Isabel's role is given short shrift. Furthermore, Maurice's supposed triumph emerges "from a state in which the sense of autonomous self is obliterated, boundaries give way to merger" ("Intimacy" 247). For Lawrence, particularly from the beginning of the war, such a merging would be anathema, but Wheeler may have hit upon the key reason for the irony in Maurice's love-in with Bertie. In short, the acute anxiety that Tate has identified in this story works against its conscious ideological scheme. Both Widmer's and Wheeler's approaches to "The Blind Man" fail to take adequate account of Lawrence's own brand of compensatory fiction. I would like to suggest a reading of the story that uses elements of both approaches, but places Isabel at the centre and takes account of Lawrence's survival anxiety.

One thing that "The Blind Man" clearly sets out to do, like so many other stories of roughly the same time, is to obliquely express Lawrence's denial of his own illness and his belief that bodily weakness is as nothing when accompanied by vitalistic strength. In this sense, the blind but potent Maurice Pervin is the chronically ill Lawrence able to overcome through force of will any bodily infirmity. However, Lawrence is also honest enough to mitigate his denial with hints of weakness. In fact, without the weakness, the denial has little purpose. The relationship of Maurice and Isabel has begun to founder on Maurice's dependency on her, an entirely reasonable state of affairs given the circumstances, but one that places him in the uncomfortable role of child or infant. One might, indeed, posit an oedipal constellation at work when Maurice, upstairs changing, hears Isabel and Bertie talking and feels "a childish sense of desolation"; he "seemed shut out – like a child that is left out." But there is more to it than that, for it is clearly a feeling of dependency rather than oedipal jealousy that distresses him: "[H]e had almost a childish nostalgia to be included in the life circle. And at the same time he was a man, dark and powerful and infuriated by his own weakness. By some fatal flaw, he could not be by himself, he had to depend on the support of another. And this very dependence enraged him" (*EmyE* 55). The narrative strategy Lawrence uses to solve the tense situation he has set up is ingenious both fictionally and psychologically. Maurice beats a tactical retreat from a contest with Bertie on Bertie's own ground, the illuminated, civilized space of intellectual intercourse. Earlier in the story, his prolonged stay in the stable prompted Isabel to enter his domain in search of him. Now, in effect, by staying away again, Maurice forces Bertie – who is, significantly, sent by Isabel to fetch her husband – onto his territory, the dark, primitive world where touch prevails.[20] In the stable, it is no contest, and the apparently disabled man wins the fight for survival and for Isabel, thereby proving his superior fitness, appearances notwithstanding. As a result, it is Bertie's self-boundaries that are destroyed: "He could not bear it that he had been touched by the blind man, his insane reserve broken in. He was like a mollusc whose shell is broken" (*EmyE* 75). In a dynamic something like that of the original function of a scapegoat, Maurice Pervin seems to have overcome his own vulnerability by "transferring" it to another person,[21] a form of physical projection that may point to Lawrence's own psychological methods of dealing with his perceived weaknesses.

There has been a good deal of speculation about the real-life model for Bertie Reid and to a lesser extent for Maurice and Isabel. Good cases have been made for both Bertrand Russell and J.M. Barrie as Bertie's model, while J.M. Murry seems to be the dark horse candidate.[22] It is instructive to recall, however, that, for all his bitterness at Russell, Lawrence did at one point admit that the vehemence of his criticism was partly the result of projecting aspects of himself onto Russell, that he too was guilty of the very things he railed about.

Approximately two months after his violently vituperative letter to Russell ("You are simply *full* of repressed desires, which have become savage and anti-social," etc. [*L* ii 392]), Lawrence wrote to his erstwhile friend, "After all, my quarrelling with you was largely a quarelling [*sic*] with something in *myself*, something I was struggling away from in myself" (*L* ii 442). Although it may be difficult to see Lawrence as either the slow-witted and heavy-limbed Maurice Pervin or the repressed and cultivated Bertie Reid, I would argue that he is both. Both the blind man and his antagonist, beyond any other identifications we can reasonably propose for them, represent aspects of Lawrence himself. This version, we must remember, was written a long time after Lawrence's feud with Bertrand Russell had turned rancid, so one would think that something other than the white heat of personal animus must be going on. Ross points to an October 1915 letter Lawrence wrote to Lady Cynthia Asquith in which he states apropos of the war, "But now I feel like a blind man who would put his eyes out rather than stand witness to a colossal and deliberate horror" ("Mythology" 304).[23] Marks suggests that Maurice is an obvious mask for a Lawrence suffering from "the self-pitying impulse to withdraw from a war-torn, censor-ridden world into an atavistically conceived and hermetically enclosed *Heimat*" (181). Delany sees the possibility of a Lawrentian self-identification with Bertie even if J.M. Barrie is the model: "[Lawrence] was repudiating a man [Barrie] whom he had once seen as a virtual double of himself" ("Who Was" 529).[24]

Lawrence was clearly much more the intellectual Bertie Reid than he cared to admit at times and, like Bertie, disliked being touched. If the name Bertie easily suggests Bertrand Russell, it also harkens back to Lawrence's own childhood name. While Lawrence in some ways would like to think of himself as a Maurice Pervin, his strength of character and vitalism overcoming mere physical handicaps, he must also fight against barely acknowledged fears of a psychological collapse like that of a Bertie Reid, fears that the hardened shell protecting the fragile sense of stand-alone selfhood will be crushed like the exoskeleton of some soft, vulnerable creature. But there is surely also a textual ambivalence vis-à-vis Maurice, who, if we use the Freudian template for a moment, is both a phallic tower of strength and a man symbolically castrated with the loss of his eyeballs, every bit as challenged a masculine creature as Bertie with his "sunken eyes" at the end (*EmyE* 63). Lawrence is using this text to deal with his own anxieties: "Lawrence uses the idea of war injury to transform the terror of castration, blindness, or any other kind of physical damage into a source of power … The historical experience of the war is simultaneously present and absent in his work, transformed into erotic fantasies which both confront and efface the damaged bodies it produced" (Tate 109).

Unlike Basil Apsley or Clifford Chatterley, other Lawrence figures wounded in the war, the blind man can reproduce. He is fit in a Lawrentian sense because his blindness is meant to make him more whole rather than less. Maurice, the blind man, personifies the metaphoric darkness where Alvina, in *The Lost Girl*, first senses salvation in the features of Alexander, her dark-visaged suitor from "down under"; his blindness approximates the literal darkness she found down under in her father's mine shaft. It is also the same lightless region evoked by the Pluto-like Count Dionys in *The Ladybird* when he tells Lady Daphne that she, like Persephone, must divide her time between being his "bride" in the nighttime world of darkness and remaining with her husband in the daylight world. In a world gone mad with mental manipulation of the material world and with products that visually titillate (such as cinema in *The Lost Girl*), blindness can be a distinct advantage in that it forces one into an immediate connection with the sensual self. It is that self that is usually salvational, the key to survival for Lawrence. Nevertheless, doubts about Maurice's strength and fitness creep into the text. The crisis in the story is triggered by Isabel's pregnancy and the fact that there will soon be a child, a blind and newly awakened creature to compete with Maurice in infantile dependency.

The opening of the story sets up a binary situation with Isabel's attention split between the absent Maurice and the expected Bertie, just as the house at the Grange farm is described as divided between front and rear, the latter the domain of the farming family, the Wernhams, and further divided between house and outhouses where the animals are kept. The structure anticipates the obvious Lawrentian contest between dark and light, lower and upper, body and intellect that is embodied in the two men and in the contrasted settings of the "hot animal life" in Maurice's stables (*EmyE* 52) over against both Bertie's "beautiful rooms overlooking the river" (*EmyE* 57) and Isabel's own well-lit and sweetly scented domestic space in the house: "Isabel was always glad when they had passed through the dividing door into their own regions of repose and beauty" (*EmyE* 53). One critic suggests that the Wernhams' abode at the back of the house acts "as an evolutionary link between the daintiness of Isabel's front parlour … and the pure irrationality, impersonality, and demonism of nature at its backward reaches" (Marks 187). What is more subtly presented and becomes clear only at the end is that the contest between the two men is not only a fight over Isabel and, as such, a proxy battle for her own internal conflict between the tendencies that the two men represent, but also a battle for emotional survival. However, ominous signs of the struggle for survival appear early on. Isabel, at times, wishes "she could be snatched away off the earth altogether, anything rather than live at this cost" (*EmyE* 47). And their first, pre-war infant has literally failed to survive. This is in spite of the fact that, even before he is blinded in battle, Maurice's lack of intellect would seem to symbolize the kind of dark world of surcharged senses that Lawrence at times argues is necessary for the overly rationalized modern world to regain some sort of balance. Life with Maurice for Isabel – and, again, she is the key character – is a soporific one of peace and maternity. Bertie represents a "re-awakening" (*EmyE* 49) and the life of the intellect. Caught between the two men and between the two tendencies in herself that they externalize, she is in a no-man's-land between warring factions that recalls the conflict that has ripped Europe apart: "Ah, why had she no peace? These two men, why did they tear at her?" (*EmyE* 50).

For Bertie, the survival threat exists externally only because of his internal weakness. He is threatened because he is a kind of hollow man who cannot withstand very much external pressure: "At the centre he felt himself neuter, nothing" (*EmyE* 58), so that, without action and thought, for him, there is literally nothing, as he suggests (*EmyE* 59). In fact, the final scene of annihilation with Maurice, when Bertie's "insane reserve" is penetrated and his protective shell is broken (*EmyE* 63), takes place precisely and ironically because Bertie's own inability to admit to his true feelings leads him to allow himself to be touched and to touch.[25] Isabel is the real loser in this reading of the text. Although she will have the child, the isolation that Maurice fears will weigh on her is now symbolically if not literally assured, for Bertie will not want to return, and with him the world of light and intellect will disappear for Isabel.

In his 1965 work, *Double Measure*, George H. Ford discussed Lawrence's extensive use of the Pluto-Persephone myth as a structure in his work: "The possible variations of a classic myth, which Lawrence saw re-enacted daily in his own home in Eastwood, provided him with a seemingly inexhaustible source of inspiration for fiction" (31), beginning with *The White Peacock* wherein Lettie complains that George Saxton is blind and unformed (Ford 36). Ford posits an S curve over the course of Lawrence's career that takes his sympathy from Persephone (essentially his upwardly aspiring mother) to Pluto (his coal-blackened father who toiled underground). Clearly, the light-versus-dark structure of "The Blind Man" suggests the Pluto-Persephone myth as an appropriate intertext. Although the title indicates that the story will focus on Maurice, the story actually begins from Isabel's point of view and seems to end with her as well; even the final sentence comparing Bertie to a crushed mollusc seems to be a free indirect rendering of Isabel's thought. One advantage to viewing the story as a variation on the Pluto-Persephone myth is that it increases our awareness of Isabel's importance in the story as the principal character[26] who is seeking a Lawrentian balance in her life (Maurice has but a vague desire to reach beyond the farm life) just as Persephone was allowed to balance her life between the upper world of Demeter, her mother, and the lower world of Pluto, her ravisher. Comparisons with *Oedipus* and *Lear* (Cushman 494), relevant as they may be to some degree, tend to obscure the importance of Isabel's role. It is not insignificant that Lawrence has suppressed, so to speak, the role of Demeter in this story or at least transformed it beyond recognition by a cross-gender manoeuvre.[27] But Persephone is very much present, and viewing the story from Isabel's perspective allows us to see more clearly what is at stake for her and whether she truly does lose in the end. The story is so odd when its apparent aim is compared to its actual shape that we might well say with Keith Cushman that it "is almost a parody of [Lawrence's] lifelong imaginative effort to integrate the powers of darkness and light" (497).

Judith Puchner Breen has written a study of "The Blind Man" that recognizes Isabel's central role but sees her as the ultimate winner in something of a contest for survival. For Breen, Isabel "ruthlessly pursues her procreative instincts" and manipulates the two men in order to get Maurice "off her hands" so that she may devote herself to the child soon to be born: "Far from being winners or losers, Bertie and Maurice are both victims of Isabel's self-interest" (64). Breen's analysis is useful in implicitly pointing to evolutionary forces at work in the story, forces that the war initially has perverted: "Only an exaggeration of the historical decadence deplored by Nietzsche and the Spenglerians, modern warfare inevitably spares weaklings like Bertie, placing in their hands the forfeited futures of the strong" (Marks 183). Now Isabel looses those same forces upon the two men, each weak in his own way, in a battle for the survival of a life not yet even born. She must be strong, even ruthless, in her project to protect the weak newborn. If Bertie knows he is beaten at the end, even devolved to an invertebrate state, the reader realizes that Maurice's weakness has done him in as well, although Maurice is blithely ignorant of the fact. In the 1920 revision, Lawrence has transformed a story about two partial men, each of whom finds added strength in allying himself to the other, into an ironic one in which the weaknesses of both are revealed. The image of the crushed mollusc Bertie, whose shell is weakened by hollowness within, is countered implicitly by the image of an Isabel solidified within by the added life she is carrying. The hollow man is opposed by the replete woman.

Although in the oedipal reading, Maurice is the child in relation to the parent figures, Isabel and Bertie, from another perspective, Bertie himself represents a child, specifically the child that Maurice and Isabel are "expecting" the way they are expecting the arrival of Bertie at the beginning of the tale.[28] Ruderman has argued that the first child dies so that Isabel can mother her wounded husband (83), but it might be more to the point to say that Bertie arrives so that Isabel can test out whether her husband can do with less mothering. By throwing the two men together, Isabel is (perhaps unconsciously) testing Maurice's reaction to the coming child. Maurice says to Bertie when he has explored the other man's head with his hands, "'Your head seems tender, as if you were young'" (*EmyE* 62). Bertie as symbolic infant in the womb space of the barn competes with Maurice for the wife/mother Isabel just as Lawrence felt he was always competing with Frieda's children for attention. While Maurice may feel a largely unconscious reassurance that his marriage can survive the child's arrival on the scene, Isabel must be horrified at Bertie's annihilation both for what it means for her friend and in that it suggests Maurice might inadvertently harm the child while trying to get to know it.

Everything Lawrence wrote during this period and the major thrust of the story itself suggests that Isabel's victory over the two men is not what he intended to portray. The overwhelming anxieties of this time, however, have created a subtext that reveals Lawrence's worries about his own survival both in terms of the intellectual self whose ego boundaries have been smashed in by the hostility of his country and of the reading public and in terms of the dark, strong, sensual self he would like to believe in but whose existence is belied by his illness and his physical decline. Although Isabel as both mother-to-be and as "nurse" is exactly the type of figure to attract Lawrence's fictional hostility, she is also in a sense a Lawrence surrogate about to give birth to a new form of fiction. She is, after all, an active writer, and the ruthlessness of her treatment of Maurice and Bertie, at least according to one reading of the text, accords well with Lawrence's postwar mood.

"HADRIAN": VICARIOUS SURVIVAL

Like "The Blind Man," "Hadrian" is, as the publisher's title "You Touched Me" indicates, a story in which touch is paramount. The sense of touch in relation to the vulnerable self can function to reinforce boundaries, that is, remind the self (and others) where it begins and where it ends, as in the poem "New Heaven and Earth," and/or pose a threat to the moat of protection around the self, as does Jim Bricknell's punch to Lilly's solar plexus in *Aaron's Rod*. In "Hadrian," the theme of touch is the "official," ideological one;[29] that is to say, it is something Lawrence believes in fervently but intellectually, and it carries with it an inversion of the Sleeping Beauty structure that is so pervasive in this period. But what drives the story emotionally and is, I would argue, the actual source of most of its power is a subliminal struggle for survival that is dramatized through a ruthless example of sexual selection. Because Lawrence's own emotions and largely unconscious fears are so heavily invested in the subliminal theme, the result is a complexity that belies the apparent simplicity of the action and produces a sense of puzzlement in the reader. Although the outline of the tale is typical enough, "Hadrian" is one of Lawrence's oddest works when we probe a little farther. The oddness derives in large part from the role of a dying old man controlling the lives of his children in a narrative that purports to be about something else entirely. Not since *The White Peacock* and George Saxton's old aunt (and Meg's grandmother), another bedridden, propertied autocrat, has there been such a figure in Lawrence, and there she was largely benign. The father-figure in "Hadrian" is a ruined shell of a man on the verge of extinction, and Hadrian, his adopted son, becomes his vehicle to symbolic, vicarious survival. The strong hint of symbolic incest in the story tends to obscure the theme of survival, but little else can explain the seemingly irrational and stubborn insistence of Ted Rockley to cut off his daughters financially if the chosen one, Matilda, does not marry Hadrian. Cowan speaks of "a kind of two-way oedipal street whereby Hadrian attains sexual conquest of his quondam 'mother,' and whereby Matilda … gains sexual access to her father through Hadrian, a younger version of himself" ("Touch" 124). This observation, I think, in spite of Cowan's reference to Ted Rockley's "collusion" in Matilda's match, understates the degree to which old Rockley insists upon the match and invests his remaining life energies in manipulating events so that it comes about, for beyond the oedipal urges that are indisputably at play is the overriding death anxiety that Rockley is struggling with. He is on a one-way street to death.

Ted Rockley is dying of kidney disease. His wife has died some years before, leaving him with the upbringing of his four daughters, two of whom are living at home as "old maids" (in their early thirties) when the story opens. Home is the Pottery House, which lies within the grounds of a now disused pottery works. Years before, Rockley, tired of having nothing but female company at home and possessed of an identification with the working class, adopted a charity boy of six by the name of Hadrian. There is a standoffishness between Hadrian and his "cousins," as they insist on being called, that makes Matilda's later touch all the more significant, but with the father there is kinship: "His Uncle, Ted Rockley, understood him tacitly, their natures were somewhat akin. Hadrian and the elderly man had a real but unemotional regard for one another" (*EmyE* 94). In the story's present, Hadrian, who left the Rockley household as a mere youth of fifteen in order to emigrate to Canada and who subsequently enlisted when war broke out, returns as a grown man for a brief visit, intending to return to Canada when his leave is up. The two daughters, Emmie and Matilda, feel that Hadrian is "sly" and suspect that he is after some kind of legacy from their father. The crucial plot device of the story is the moving of Ted Rockley's bedroom downstairs so that Hadrian can stay upstairs in Rockley's usual room. The room change is presented in a matter-of-fact manner that reflects arrangements that seem practical enough in terms of caring for a dying parent, but the obvious symbolic implications of lodging the aspiring "son" in the father's room are not lost on Lawrence. During the night, in an odd somnambulistic state, Matilda goes into her father's room, "forgetting" that it is Hadrian who is sleeping there, and she strokes his forehead until, when Hadrian speaks to her, she realizes her mistake and bolts from the room. The touch gives Hadrian the idea that he would like to marry Matilda, ten or eleven years his senior, and he broaches the subject with Rockley. The old man is at first shocked at the idea, but he soon warms to it, almost literally: "He had flushed, and looked suddenly more alive. Secretly he loved the boy" (*EmyE* 101). He warms to it so much that he changes his will so that Emmie and Matilda are cut out of it altogether and Hadrian is left the entire estate unless Matilda consents to marry her "cousin." Emmie tries to frighten and shame Hadrian out of his scheme, but Hadrian will not be balked. He insists to himself that he does not want Matilda for the money, although he does want the money badly enough. Initially horrified at the idea of marrying Hadrian, Matilda suddenly gives way and agrees after he explains to her that, although he realizes that her touching him was a mistake, he cannot forget it. After the marriage, Matilda and Hadrian rush back to the dying man, who says enigmatically to Hadrian "'I'm glad you're mine,'" and then requests in a voice that is "strange and unrecognizable" a rare kiss from Matilda, who is then to kiss Hadrian in what seems to be a union of the two men through the daughter/wife (*EmyE* 107).

A number of commentators have noted the many similarities between "Hadrian" and *The Fox*, particularly the situation where a younger man of a lower class just back from the war selects one of two older women as a marriage partner and then ruthlessly proceeds to manipulate events so that he gets his way.[30] In both cases, property or money seems a secondary but distinct added motive to the marriage proposal, at least initially. As a matter of fact, the two stories were written at about the same time. An earlier version of *The Fox* was sold in June 1919 to *Hutchinson's Magazine*, the same month that Lawrence began writing "Hadrian." The latter was published as "You Touched Me" in *Land and Water* in April 1920 and was slightly revised for publication in *England, My England* at the end of 1921, just when Lawrence was transforming his short story about the fox into novella length. There are also many features in the two works, of course, that distinguish them. In "Hadrian," there is the crucial importance of Ted Rockley, the dying father, and the inadvertent touch around which the plot seems to coalesce.

The importance of touch in "Hadrian" has been dealt with most notably by Cowan. In an essay entitled simply "Lawrence and Touch," Cowan succinctly articulates the manifest intention of the short story: "Because its motivations are largely unconscious, touch has an extraordinary power to bind relationships and to bring into consciousness hidden forces at work in them. 'You Touched Me' is a fable about the power of a single touch to be irrevocably binding" (123).[31] However, as we have seen with other works from this period, there is a different, less obvious intention at work in "Hadrian" as well, one that is related to Lawrence's survival anxiety and that takes the form of a resolution in terms of the human interaction with the forces of natural and sexual selection. Keith Cushman, for one, senses something odd working underneath the obvious storyline: "Lawrence once again seems to be dramatizing a belief in the primacy of the senses, and once again closer inspection reveals otherwise." For Cushman, the underlying drive is the desire to illustrate "the fundamental hostility between men and women" (499). There is certainly a gender hostility in the story, for instance, when the narrator tells us that Ted Rockley, in conniving with Hadrian's plans, "seemed to have a strange desire, quite unreasonable, for revenge upon the women who had surrounded him for so long, and served him so carefully" (*EmyE* 104) as if those tending to him had caused the illness. However, what I would like to argue is that such enmity is part of a larger psychological constellation that creates the subtext of survival. Lawrence is interested, in a mostly unconscious way, in exploring other issues than the power of touch.

A hint of something different is in the story at the very beginning, where we meet a moribund pottery works and a decrepit man that seem to be versions of each other. The senses of finality and confinement are palpable on the first page:

> The Pottery House was a square, ugly, brick house girt in by the wall that enclosed the whole grounds of the pottery itself … The pottery itself was now closed, the great doors of the yard permanently shut. No more the great crates, with yellow straw showing through, stood in stacks by the packing shed. No more the drays drawn by great horses rolled down the hill with a high load. No more the pottery-lasses in their clay-coloured overalls … All that was over. (*EmyE* 92)

The pottery works are an emblem of the physical decline of Ted Rockley, whose life now too is about to be "no more." But they are also symbolic of the kind of "moulding in clay" that Rockley undertakes in trying to remake himself through the adoption of Hadrian, who becomes Rockley's "double" but with the essential difference that the younger man, despite Emmie's angry declaration to Hadrian that "'you're not fit to live'" (104), is fit precisely to live, fit for survival; in fact, from the perspective of this reading of the tale, Emmie's comment is significantly ironic. That Matilda forgets that Hadrian is sleeping in her father's room – on one level, the barest plot pretext for the act of touching that Lawrence tries to make the centre of the story – reinforces Hadrian's function as a stand-in for Ted Rockley. The final scene of exchanged kisses underscores both the incest theme and the double function Hadrian serves for the dying old man, whose vicarious enjoyment of Hadrian's triumph is also a victory of sorts for himself in terms of continuing life.

F.R. Leavis tended to see Lawrence as a kind of anti-Victorian hero of vitalism who stood for life as the central value in a society that in various ways worked to vitiate it. Such was not an unreasonable view of Lawrence in the first decade or two after the unimaginable destruction of the Second World War, especially given Lawrence's prediction that, as long as his culture refused to come to terms with its deathly imbalance, the business begun with the First World War was not finished. In his poem "Noli Me Tangere" (from *Pansies*), Lawrence explicitly though negatively links the power of touch with his culture's imbalance: "Great is my need to be chaste / and apart, in this cerebral age. / Great is my need to be untouched, / untouched"

(*Poems* 469). In *D.H. Lawrence: Novelist*, Leavis, although he does not quite endorse the ideological position of "Hadrian," does argue that Ted Rockley's motives in compelling his daughter to marry Hadrian are in keeping with Lawrentian vitalism: "In the father's confident sense of things, Hadrian stands for life, and the marriage is the assurance of a living future; ... essentially the dying man sees the dead-end of old-maidhood in the Pottery House as a defeat of life" (307–8).[32] Such a reading, however, cannot adequately take account of the disproportionate emotional reactions in the story, Hadrian's to the mistaken touch – an absurdity from a realistic point of view – and Rockley's at the prospect of Hadrian marrying Matilda. If, however, we view Rockley as personifying Lawrence's deepest fears about his own robustness and Hadrian his hopes for survival, then the exaggerated responses make some sense.

Embedded in "Hadrian," as in so many of the other works of fiction of this period, is the Sleeping Beauty story, which Freudian critics such as Bruno Bettelheim have interpreted as an expression of the growing into sexual maturity of a young woman (the bleeding from pricking the spindle), the latency period before an emotional readiness for sexual relations (the one hundred years' sleep), and the initiation into adult sexuality (the awakening kiss that coincides with the passing of the allotted time) (Bettelheim 225–36). Lawrence incorporated his own version of the fairy tale into his personal mythological system, according to which we usually have a woman, sleepwalking through life, being awakened to a fuller, vital existence by the presence of the right man. In "Hadrian," there are various forms of awakening. The most obvious inverts the usual gender roles in that it is Hadrian who is awakened by Matilda's touch into an awareness of life's possibilities if only he will seize his opportunities. Matilda would like him to go back to sleep, so to speak, but, unlike the female Sleeping Beauties in the fiction of the period, who are urged to return to their passive ways of life, Hadrian refuses to resume his slumber: "'If you wake a man up, he can't go to sleep again because he's told to'" (*EmyE* 106). Ted Rockley is the intended recipient of Matilda's awakening caress, but the intention itself is rather hazy. The confusion of her father and Hadrian occurs shortly after Rockley has told Matilda that he has left everything to Emmie and herself (the two other daughters seem to have been forgotten), and that he wishes them to give Hadrian his watch and chain and one hundred pounds (*EmyE* 98). In an "entranced" state that night, "she thought of her father, only her father. At last she felt she must go to him" (*EmyE* 99). But it is Hadrian she goes to, thus indicating that her unconscious, perhaps prodded by the thought of her father's wish that those symbols of masculine mastery, his watch and chain, pass to the younger man, has led her to the conclusion that Hadrian is the appropriate person to replace her father as head of the Pottery House. As for Rockley himself, he is temporarily revived by the idea of his daughter marrying his adopted son. Feeling "a new access of strength," his face regains "some of its old, bright handsomeness" (*EmyE* 104).

The final Sleeping Beauty is Matilda herself. Harris is right to point out that no one touches Matilda and that, on the "quotidian level, [Hadrian] has not awakened Matilda, he has forced her" (152). Nevertheless, she is woken up. The narrator makes it clear that she enters Hadrian's room in a state barely more conscious than sleepwalking and that the unexpectedly fresh and smooth feel of the face she assumes to be her father's surprises her "in her entranced state. But it could not waken her" (*EmyE* 99). Only Hadrian's voice, like that of several other male Prince Charmings of the period, does the trick: "It was Hadrian's voice. She started away. Instantly, she was wakened from her late-at-night trance. She remembered that her father was downstairs, that Hadrian had his room" (*EmyE* 99). The simplest interpretation of Matilda as a Sleeping Beauty is that she too has been awakened to life's possibilities by her touch of Hadrian and by his inordinate reaction to it, after having resigned herself to a spinster existence in a town where few eligible bachelors of the right class are available: "[B]ank-clerks or nonconformist clergymen or even schoolteachers having failed to come forward, Matilda had begun to give up all idea of ever leaving the Pottery House" (*EmyE* 93). However, in this case, the simplest explanation will not do. Even though the Sleeping Beauty paradigm has and always will have a powerful connection for Lawrence with his own and Frieda's early life together, at this point in his career the female Sleeping Beauty figures seem much more complicated than the male. A pertinent question is this: to what exactly is Matilda awakened in the end? To a sexually active existence, even children, perhaps, but it remains unclear finally if she will ever leave the confines of the "ugly brick house girt in by the wall that enclosed the whole grounds" (*EmyE* 92). Matilda may be more Rapunzel than Sleeping Beauty. As Widmer suggests, "[t]he power of physical contact in the unconscious – the fairy-tale motif of the awakened sleeper – traps Matilda" (58).

Images of confinement in the story, in fact, seem particularly to be associated with the two women. Rockley is largely confined to his bed, but he habitually looks out upon his beloved garden, seeking perhaps for inspiration about how to create an Edenic constellation of relationships after he is gone or, it may be, contemplating the means by which he may regain the immortal state that was humanity's before the Fall. Hadrian has the air of one who comes and goes as he pleases and is on his way to Canada. The women are the ones who seem enclosed in several ways. When Matilda first learns the terms of her father's revised will "[s]he rose and went out of the room, up to her own room, where she locked herself in" (*EmyE* 103). The male control of the women in the story – unusual in the pre-*Plumed Serpent* Lawrence – is an expression of both Lawrence's need to alleviate his own fear of confinement and his association of women with the forces that seem to threaten his selfhood and his existence. Matilda's choice in this story of "harshness, compulsion, and cruelty" (Harris 151) is between a marriage she appears not to want but that will allow her to keep her share of the property (for Hadrian, it turns out, will get nothing according to the terms of the new will if Matilda consents to marry him) and penury and no prospect of marriage. Her decision finally to marry Hadrian is not explained. She simply informs her father that she is now willing to accede to his demand.

Matilda is also awakened to the fact that her father feels tyrannical malevolence instead of gratitude for her years of care, but her awakening seems ill-suited to the Sleeping Beauty tale that is so clearly present in an intertextual way. Sleeping Beauty's long sleep, a sleep that encompasses the entire kingdom in Grimm, is also an expression of Lawrence's horror of fixity and ultimately death. The princess in the fairy tale enters a petrified, fixed state; she does not change at all in one hundred years. Her awakening is a kind of resurrection. All three characters in "Hadrian" face different versions of such states. Hadrian seems fixed in class hatred and self-sufficiency; Matilda is set to enter a long period of spinsterhood; and Rockley is on the edge of extinction. The marriage of Hadrian and Matilda appears designed to deal with these rigidities and with the underlying problem of survival. Matilda's sexual selection dilemma is resolved; Hadrian will survive economically (as he survived the war physically); and Rockley will survive vicariously through Hadrian. Leavis is ultimately right that the energy in the story is directed against deathliness and toward life, but the issue is far more complicated than Leavis's reading would suggest.

As both Maurice and Bertie played out aspects of Lawrence's anxieties in "The Blind Man," here both Ted Rockley and Hadrian embody aspects of Lawrence's psyche. The old man is the sick and dying Lawrence, the Lawrence who identified with his father and who longed at times for a child. Hadrian is the vitally alive Lawrence, the cunning working-class lad who always managed to survive, who married an older woman of a superior class and who wished to start a new life abroad. Ruderman points to a February 1920 letter by

Lawrence in which he says, "I am a sort of charity-boy of literature, apparently. One is denied one's just rights, and then insulted with charity" (82).[33] The pre-oedipal concerns that Ruderman's study so ably delineates in Lawrence's work are significant in "Hadrian" as well, but there is more going on here than the split male character conspiring as one to take revenge upon the "devouring" women. Rockley's "I'm glad you're mine" directed at Hadrian – a mere pronominal form removed from the brutally frank but impossible "I'm glad you're me"– and the pointedly significant kiss between the father and daughter resonate as expressions of the desire on the part of Rockley to live on by proxy through the younger man. Another token of the desire to survive in some sense is Rockley's strong wish that his daughters remain at the Pottery House, a place with which he identifies strongly, after he is gone (*EmyE* 98). For his part, Hadrian feels that he is uniting with Rockley somehow in marrying his daughter, a clear manifestation of sexual selection: "The same glamour that he knew in the elderly man he now saw in the woman. And he wanted to possess himself of it, he wanted to make himself master of it" (*EmyE* 100). Ted Rockley will die, but he will perish knowing that he lives on through a young, robust man who, as a child, was selected by him and has been moulded, as a clay pot is moulded, into a young version of the man he himself would want to be. Lawrence can fantasize his own continuing existence by creating such characters, and, just as the totemic fox of the story of the same name *was* that story too, so the eponymous character Hadrian represents the fiction Lawrence wishes to write, strong, ruthless and clever.[34] He will replace the old, now-moribund form of writing that cannot survive the postwar world with a new fiction that will bend the marriage fiction, "The Sisters," to its will just as Hadrian triumphs over the two sisters Emmie and Matilda. It is a fiction that looks to an overseas audience as Hadrian the character looks to Canada.

"THE HORSE-DEALER'S DAUGHTER": THE SICK PHYSICIAN

The first version of the "The Horse-Dealer's Daughter," then called "The Miracle," was written in the middle of the war, and the revised version done at the same time that Lawrence was revisiting the war in much of his other fiction of the period; nevertheless, the war setting is absent, as is any overt battle of the sexes or struggle for mastery between men and women. The story is about the Pervin family (the same surname as in "The Blind Man"). The father has died, leaving crushing debts that will ensure that the horse-dealing business he established must be wound up and the home, Oldmeadow, vacated. The three brothers, Joe, Fred Henry, and Malcolm, and the one sister still at home, Mabel, watch as the horses "tied head to tail" are taken out for exercise by the groom for the last time (*EmyE* 138). The brothers have various plans for their future, but Mabel, who has kept house for the family since her mother died thirteen years earlier, refuses to answer the question that opens the story: "'Well, Mabel, and what are you going to do with yourself?'" (*EmyE* 137). As we eventually find out, the reason for Mabel's sullen silence is that what she plans to do with herself is to do away with herself. She is determined to join her mother in "the world of death" (*EmyE* 143) through suicide. The brothers assume that Mabel will lodge temporarily with their married sister, Lucy, and the family confabulation is about to break up when Jack Fergusson, the local medical assistant and particular friend of Fred Henry, walks in to say his good-byes and to plan a final evening with his friend. It is a gloomy winter's day, and the doctor, who is suffering from a heavy chest cold, is bundled up against the chilly air. Later that day, Fergusson passes by Mabel while she lovingly attends to her mother's grave, and, later still, at dusk, he spies her walking toward a pond that lies below Oldmeadow. He sees her walk deliberately into the water until she disappears under it, at which point he rushes to her aid. Fergusson is able to save Mabel's life, but he nearly drowns in the effort, being a non-swimmer. He laboriously carries her back to Oldmeadow, where he strips off her sodden clothes, wraps her in a blanket, and lays her down in the kitchen by the fire that is burning in the grate. He is about to attempt to find some dry clothes for himself as "he was mortally afraid for his own health" (*EmyE* 147) when Mabel comes to. She forces Fergusson out of his professional and distancing medical role and elicits a declaration of love from him and a promise of imminent marriage.

Of all the stories of the *England, My England* volume, "The Horse-Dealer's Daughter" would seem to harken back most emphatically to the erotic explorations of *The Rainbow* and the marriage period of Lawrence's writing in general. The story, as we have seen, ends with the prospect of marriage between Jack and Mabel, but Lawrence has planted the seeds of doubt regarding matrimony earlier in the story, when we learn that Joe's plans, ironically, involve emulating the horses he has heretofore been master of: "[h]e would marry and go into harness. His life was over, he would be a subject animal now" (*EmyE* 138). Furthermore, the end of the horse-dealing business with which the story begins and, in particular, the "cavalcade of shire horses swinging out of their own yard" that are to be "the last horses" the Pervin brothers will deal with (*EmyE* 137) may be a metafictional signal that Lawrence is done now with the kind of exploration he undertook in *The Rainbow*. We have seen a repudiation of the rainbow symbol itself in *The Fox* and in *The Lost Girl*. It now appears that the horses that loomed so significantly during Ursula's crisis at the end of the earlier novel are being put out to pasture, so to speak, in the Lawrentian canon.[35] Far from being the symbols of instinctual life that the horses were in *The Rainbow* (and will be again in *St Mawr*), the animals here are associated with the rather stupid, languorous aimlessness of the Pervin brothers that Mabel detests: "Every movement [of the horses] showed a massive, slumbrous strength, and a stupidity which held them in subjection" (*EmyE* 138).

Harris's reading of "The Horse-Dealer's Daughter" is, as usual, highly competent (although she calls Jack Fergusson "Sam" throughout). It may serve as the articulation of the general consensus that has coalesced over the years around a story that has attracted much critical attention. Harris calls "The Horse-Dealer's Daughter" "perhaps Lawrence's best-known tale of resurrection." It is a "ritual progress out of superficiality into death and rebirth" (125).[36] Most particularly, this is a story of "a dual rescue," since Dr Fergusson feels enslaved by his work and attempts to escape his undesirable existence in a manner that is just as deathly as Mabel's obsessive thoughts of her deceased mother (126).[37] The pond scene is a symbolic baptism for both characters into a new life as well as a symbolic delivery of a newborn life by the doctor (127).[38] Mabel's corresponding "delivery" of Jack occurs in the kitchen when she struggles to break through his professional reserve and dredge to the surface his repressed affective self (127–8). In conclusion, according to Harris, this story "dramatizes its male and female characters with a fine tolerance for and recognition of their mutual needs, powers, dreams" (129). As Jack Stewart has pointed out, Lawrence's use of the four elements, particularly his opposing the elements of water and earth over against fire and air, functions to underscore the positive movement that has occurred in the narrative from death to life: "After the crisis of immersion, repression is overcome and the swing to Eros is symbolized by images of air and fire" (522–3).

There is no doubt at all that Lawrence intended "The Horse-Dealer's Daughter" to dramatize a movement from death toward life. The feeling of deathliness is pervasive for most of the story, encompassing the end of the horse-dealing business and the dead winter landscape.[39] Even the dining-room in the Pervin home "looked as if it were waiting to be done away with"; Mabel sits "like one condemned, at the head of the table"; and the day is "grey, deadened, and wintry" and the town full of "extinct houses" (*EmyE* 137, 140,

144). The water in the pond in which Mabel attempts to end her life is not only deadly but is itself dead, consisting as it does of gloomy "dead water" (*EmyE* 145). The countervailing force of life begins with a look that passes between Jack and Mabel when she is at her mother's grave, a glance that makes him feel as if "life came back into him, he felt delivered from his own fretted, daily self" (*EmyE* 144). When he spots her distant figure on the way to the pond, the doctor's "mind suddenly became alive and attentive" (*EmyE* 145). As he drags her from the pond and works to save her, Jack can "feel her live beneath his hands" (*EmyE* 146) while he himself, shuddering in the Oldmeadow kitchen "like one sick," needs only a look from Mabel and "his life came back in him, dark and unknowing, but strong again" (*EmyE* 147). And so on it goes in this vein; the conventional reading of the story reflects what Lawrence has carefully plotted in the work. However, things are seldom quite so simple in the narratives of this transitional period, and, in this particular case, a subtle, qualifying subtext pits Lawrence's survival anxieties associated with women against the story of a double resurrection.

The central problem with the conventional reading of the story in my view is that it fails to take sufficient account of a striking similarity between the description of the scene where the doctor nearly drowns in attempting to save Mabel and the account, later in the story, of what is normally taken to be the culmination of the theme of "mutual regeneration" between Mabel Pervin and Dr Fergusson (Stewart 521). During the rescue, Fergusson sinks into the clay bottom of the pond, "and the water clasped dead cold round his legs." He smells "the cold, rotten clay that fouled up into the water," but he persists in his rescue despite his disgust, until the water "rose over his thighs, over his loins, upon his abdomen" (*EmyE* 145). Eventually, he locates Mabel under the water and lifts her onto the bank. After the rescue, Mabel embraces Fergusson's legs, "pressing her breasts against his knees and thighs, clutching him with strange, convulsive certainty, pressing his thighs against her," and he initially recoils in horror: "It was horrible to have her there embracing his knees. It was horrible. He revolted from it violently" (*EmyE* 148). Both descriptions are sexualized ones involving Fergusson's legs, and both express abhorrence on the doctor's part. It is as if Mabel becomes the personification of the muddy water that gripped his thighs during the rescue and threatened him with death, and one is reminded of Pinkney's description of Lawrence's view of female sexuality as something that entices "a reluctant but powerless male into its soft, deadly depths" (118). Fergusson watches "the strange water rise in her eyes" after he has saved Mabel, as if he were back in the pond sinking deeper into the water. Those eyes are also "wistful and unfathomable" (*EmyE* 149) as if they remind Jack of his difficulty in fathoming the depth of the pond. What are we to make of such a congruence, where the fatal, rather than the baptismal, aspects of the water are emphasized in the parallel between the water and Mabel? It would seem to run counter to the received reading of the story as one of the salvational and transformational power of eros. It is not that the similarity between the two scenes has never been noted before. Steven Phillips, for one, briefly discusses the "structural repetition" involved, where "the second descent and rebirth reinforc[es] the first" (96), but he seems to have no problem with Fergusson's reaction of horror at the cold, slimy, and smelly water. From the point of view of narrative strategy, the obvious but seldom asked question is this: if Lawrence intends this to be a simple story of a double rebirth through symbolic baptism, why make the baptismal water so decidedly disgusting?

It is just possible to see the doctor's horror in a positive light. According to Schapiro, both Mabel and Jack are leading "inauthentic" lives, and Lawrence's imagery strengthens the parallels between their defences: "As Mabel has erected a dense wall of emotional reserve, the doctor has muffled and suppressed his sensual, feeling life beneath the protective garb of his professional identity" (74). Schapiro is undoubtedly correct that the manner of dressing himself all "muffled up in overcoat and a purple woollen scarf, and his tweed cap, which he did not remove, ... pulled down on his head" (*EmyE* 140) represents something about the doctor's attitude to the world. However, we should also remember that Lawrence makes a point several times about the chest cold he is suffering from, and Fergusson's muffling up is also an effort to keep himself from getting any sicker, an effort that Mabel's suicide attempt, which obliges him to unmuffle and to get soaked in the middle of winter, works to frustrate.

Cushman, again, is one of the few commentators to suggest that "there is something troubling about Mabel Pervin's rebirth: the story is more ambiguous than a Jungian reading allows" (498).[40] Cushman sees no irony in the story (498), whereas it seems to me that there is a palpable irony present in the striking similarity between Mabel Pervin's supposedly salvational clinging to Jack Fergusson's legs and the earlier clinging water that threatens to pull him down to his death (and thereby, *a fortiori*, also ensure hers). One could argue, nevertheless, that, just as Mabel must enter the watery element to be reborn, so must Fergusson "drown" in Mabel's eyes in order for him to be resurrected. Widmer suggests that the story portrays "love as itself a kind of death" (173). The problem with this is that the reader can only with great difficulty accept Mabel as the agent of both actual death and of symbolic rebirth at the same time. And yet this contradiction accurately reflects the ambivalence toward women that drives this period of Lawrence's writing: a classic approach-avoidance conflict. There is a powerful feeling of dependence that propels one toward the source of nurturing and comfort coupled with a resentment and repulsion because that very dependence threatens to swallow one up. The only conclusion one can draw is that "The Horse-Dealer's Daughter," like the other stories of *England, My England*, contains a self-subverting level in which Lawrence defends against his deepest fears regarding survival, illness, and confinement by projecting them onto women. Stewart refers to the "painful loss of ego" that is necessary in the death-and-rebirth process that the story enacts (522). But we have seen in all the works of the transitional period that it is precisely protecting the boundaries of the vulnerable ego in which so much of the narrative energy in Lawrence's fictions is invested and toward which so many of the misogynist defences are directed. If the conventional reading is deficient, then how do we read "The Horse-Dealer's Daughter" and in particular its apparent approval of a ritual death-rebirth transformation through an erotic recognition between a man and a woman?

The title of the story suggests that Mabel is to be the centre of interest in the story; however, few readers would disagree that the title *Women in Love*, for example, is anything but an accurate indication of the focus of the novel. In this case, we must, I think, attempt to see the short story from the perspective of the doctor if only because his point of view is seldom considered. His chest cold, which the reader is never allowed to forget about for very long, is an obvious point of identification between the character and his creator, the chronically coughing D.H. Lawrence.[41] Lawrence may be working out in this tale his aggressive fantasy that "saving" Frieda has endangered his own health to the point where he fears for his survival. Beneath the surface of "The Horse-Dealer's Daughter," as beneath the water of the Oldmeadow pond, there is a struggle for survival that is different from the survival implicit in rebirth. If the suppressed goal of Dr Fergusson is intimacy with one of the brothers (as male bonding is central in "Hadrian," "Monkey Nuts," "The Blind Man," and "Wintry Peacock" in the same volume of stories), then Mabel's suicide attempt comes between them. There will be no matey pub crawl this night. Indeed, in finally giving Fergusson the clothes of one of her brothers (or possibly of her father) to wear instead of his own soaking garments, Mabel is both providing a symbolic threat to Jack's own identity and unconsciously acting out an incestuous desire. An unconscious allusion to the mutual drowning of brother and sister in George Eliot's *The Mill on the Floss* cannot be entirely discounted. In terms of Lawrence's fiction of the period, Mabel is the sleepwalking female (March, Daphne, Hannele, Alvina) and the obstructing one (Banford, Mrs Hepburn, Miss Frost) rolled into one. Since the doctor has a bad cold and Mabel's implicit demand on him forces him into the water and prevents him for some time from removing his wet clothes after the fact, we are seeing an unconscious symbolic enactment of the survival anxiety made acute by female domination. In effect, Mabel uses Hadrian's argument, "you touched me," to coerce Jack into a declaration of love almost as surely as Hadrian coerces Matilda into marriage.

As in so many of the other narratives we have examined, "The Horse-Dealer's Daughter" exploits mythic and folkloric intertexts.[42] As a variation of the Pluto-Persephone myth that we have seen at work a number of times in the fiction of the period, the story inverts the role of mother (Demeter) and lover (Pluto).[43] Here, Mabel, the Persephone figure, wishes to join her mother in the underworld realm of the dead and descends in order to do so. Jack Fergusson seems to cut a poor figure as Pluto, but, as Ryals points out, the story does feature imagery of hellishness (155), particularly with regards to Jack's relationship with Fred Henry: "'Well, this is the devil's own,' exclaimed the doctor when he was left alone with Fred Henry." Perdition for Jack Fergusson seems to be the departure of his male friend, whom he will miss "like Hell" (*EmyE* 141). Lawrence often applies a Blakean inversion to conventional narrative trajectories since, like Blake, he felt that his culture had a perverse and inverted value system. So on the obvious plot level of "The Horse-Dealer's Daughter," Mabel's death drive is converted through the power of eros into a renewed desire for life. She is hauled up from the underworld away from her mother by the male who is to be her mate. If we recall Lawrence's resentment against mothers, however, the narrative strategy of inverting the Pluto-Persephone story takes on a more ominous cast. Seen thus, it is the Persephone figure who threatens to pull Pluto down to his death and away from his true love, the brother, Fred Henry. In fact, notwithstanding Mabel's supposedly life-giving gaze, Jack is a very unwilling lover. Most readers of the story would no doubt argue that that is because passion must gradually break through the disinclination of the will to let go of everything in the conscious mind in order to be saved in a Lawrentian manner. And yet the language describing Jack's reluctance is strong, as if Lawrence's own resentment were colouring his conscious design for the story: "It was horrible to have her there embracing his knees.

It was horrible. He revolted from it violently" (*EmyE* 148). Even after he has committed himself to Mabel, Jack's kiss is "half in anger" (*EmyE* 151). In this inverted version of Pluto and Persephone, the woman's connection to her mother is deadly, while the man means life to her. However, the cost to the man himself is insisted upon in a way it never was in the marriage fiction.

The story of Sleeping Beauty is inverted in Lawrence's tale as well. We might, of course, view Jack as the Prince Charming who resurrects the slumbering princess with a kiss and, like the prince in the Grimm version of the story, happens to be at the right place at the right time to awaken the seemingly dead female. Widmer calls it "a fairy-tale coincidence" (173). It is also possible to draw a parallel between the fairy-tale prince who penetrates the heretofore deadly growth of thorns surrounding Sleeping Beauty's castle and Jack Fergusson, whose gaze penetrates Mabel's moat of safety in the churchyard: "There she always felt secure, as if no one could see her … she felt immune from the world, reserved within the thick churchyard wall as in another country" (*EmyE* 143). But even in this parallel between fairy tale and short story, it is the man who seems to be the victim of a spell rather than the woman: "He slowed down as he walked, watching her as if spellbound" (*EmyE* 144). Jack is the Sleeping Beauty in more ways than one, and this time it is the awakened male who is threatened with slumber, a slumber that paradoxically seems to be a direct result of the awakening. We have seen the connection between the Pervin brothers and the great, "slumbrous" draught horses who move "in a motion like sleep" (*EmyE* 138). What we have at the end is a Jack Fergusson, dressed in Pervin clothing, pledged to the Pervin daughter in a state where "his will seemed to have gone to sleep" (*EmyE* 147).

In "The Horse-Dealer's Daughter," Lawrence is subtly giving voice to his highly ambivalent view of women in the pre-leadership period and thereby giving us a clue as to why the leadership period came into being in the first place. Both the muddy water of the pond and, later, the mud-soaked Mabel cling to the legs of the doctor who saves her from suicide by drowning. Although the mud recalls the Brangwen soil in *The Rainbow* that clung to the feet of the Brangwen men "with a weight that pulled like desire" (*R* 10), the soil that is associated with the water and with Mabel now threatens to drag the man down to death. Like the sick physician in the story, Lawrence himself is a gravely ill man who is attempting to diagnose the malady of his culture. He allows himself to be pulled back to the arena of his greatest fictional triumphs, the love relationship between men and women, but with the objective not of repeating his ablutions as a priest of love in the pond below Oldmeadow but of working out survival anxieties that have reached a point of acuteness. There will be no miracle in the lives of the horse-dealer's daughter and the doctor, as Lawrence's change of title perhaps indicates, nor will there be a miraculous rediscovery of the pre-war vision in Lawrence's writing. From *Kangaroo* until *Lady Chatterley's Lover* (and, arguably, from *Aaron's Rod* onward), Lawrence's male protagonists will have given up on heterosexual love in favour of a male bonding that potentially encourages a rabid misogyny.

Conclusion
The Writer as Gamekeeper

In the end, of course, Lawrence did not survive beyond middle age; he was not terribly fit in the Darwinian sense. At his death, however, he faced his failure to survive bravely, "dying game" in the words of David Ellis's biographical title (taken from a Lawrence letter). He lost out in terms of sexual selection as well in the sense that, although he mated, he did not pass on his genetic material to the next generation, almost certainly because of illness-induced sterility. Lawrence knew that he was not to be lucky in the lottery of natural and sexual selection, and the force of his numerous declamations against evolutionary doctrine was a function of that realization. That he turned to that same set of ideas at a time of crisis for his health, his marriage, and his career may be ironic, but it is also understandable as a defensive manœuvre.

The phrase "dying game" reminds us that the game gamekeepers keep is for the blood sport of human amusement, an amusement that most of the landowners in England took very seriously up until recently, and still do in some cases. For the poachers, it was not so much sport as survival that was at stake for the most part. Lawrence's gamekeepers are a nice metaphor for his own ambivalent attitude to human intervention in the natural world, the survival-of-the-fittest struggle that so fascinated and repelled him, and for the role of the writer who represents that world in fiction with characters who may embody ideas that will either survive or not. The keepers partake of the struggle but also intervene in it, deciding who will live and die: for all their closeness to the natural world, they do not let nature take its course. In that sense, they are a cultural imposition on nature. In the end, the culture they embody most significantly for Lawrence is his own writing, and what they tend are his wild, unruly thoughts. Lawrence returned again and again to the keeper figure or a variant of the keeper throughout his career. In fact, keeper figures act as bookends of a sort for Lawrence's novels, featured as they are in the first novel, *The White Peacock* (1911), and the last, *Lady Chatterley's Lover* (1928).

While J.M. Murry was mostly right when he claimed in *Son of Woman* (1931) shortly after Lawrence's death that the gamekeeper figures in Lawrence's first and last novels are essentially one and the same,[1] he ignored the crucial fact of their cultural significance. Murry does recognize one important difference between the two gamekeepers, or at least in their representative weight. In *The White Peacock*, Cyril is there to offset with a spiritual self the animal self that Annable represents. In *Lady Chatterley*, Cyril, in a sense, is swallowed up, and all we have left is the animal-man, Mellors, a kind of Lawrence fantasy-self. But Murry understandably fails to see that Clifford Chatterley is also Lawrence, his diseased, disabled self. As Ellis writes, "If Lawrence did believe that he shared Clifford's plight, he was determined not to make that a reason for the slightest indulgence in his fictional treatment" (327).[2] The parallel with Lawrence's treatment of Egbert in "England, My England" is instructive. Through his narratological and psychological struggles in the transitional fiction, Lawrence has learned how to hive off his own "unfit" characteristics or what he sees as undesirable traits in others, and give them to characters who are then demonized and/or eliminated. The actual character-splitting works to some extent, as it does up to a point in Aaron and Lilly of *Aaron's Rod*, although the inarticulate Parkin of the first two versions of *Lady Chatterley* may be more believable than the Mellors who shares his creator's concern with words. The ruthlessness involved in the demonization or elimination, however, is another matter.

A second crucial difference between Lawrence's first and last gamekeeper figures is also relevant to this study of Lawrence and survival: Mellors survives while Annable does not. As Murry points out (45), Annable's death coincides with the news of the death of his former wife, the Lady Crystabel. He cannot survive the dredging up of the long-suppressed feelings concerned with her, and the cause of death, a stone wall in a quarry coming down on top of him, is symbolically appropriate. It is his own quarrying of the unconscious that Annable fears, and strengthening the gamekeeper figure through a kind of purging of "soft" characteristics so that, in effect, he may survive becomes one of Lawrence's tasks, just as in his life he had to harden himself to survive physically and to coarsen aspects of his own nature in order, as he would have seen it, to embark on his task of helping his culture survive. The stone of the quarry must be internalized, as it were, and a wall constructed on the inside around the vulnerable ego in order to protect it against the Lady Crystabels of the world. The hardness is not so evident in Mellors as it is in Parkin of *The First Lady Chatterley*, for Lawrence's fear has begun to soften by this time into a "tenderness."

That Lawrence had a grudge against women has been a critical commonplace for a long time now, but to ascribe the shortcomings of Lawrence's post-*Women in Love* fictional output only to that knotty issue would simplify a rather complex phenomenon. It is certainly deeply involved in the deterioration of the quality of his work, but, as I have argued, the causal link is not a direct one. He adored his mother as a boy, was intensely infatuated with Frieda early in their relationship ("She is ripping – she's the finest woman I've ever met" [*L* i 384]), had very close friendships with numerous women all his life, and could write with great sensitivity from a woman's point of view when he wished to (Gertrude Morel, Lydia Lensky, Anna Brangwen, Ursula Brangwen, Alvina Houghton, and even Kate Leslie). It was only when he felt his survival as man and artist at stake that Lawrence lashed out at women as the source of his suffocation and only then that he utilized in his own fiction the very force threatening him with extinction as an unfit man and writer. Darwinism in the transitional fiction is the intervening factor in Lawrence's development as a writer. His ruthless handling of female characters in the transitional and leadership fiction is, as it turns out, not only an artistic misjudgment but also a symptom of a man using as a fictional tool a force whose truth he has denied in order to try to shed in his books the sicknesses besetting him. The fiction becomes duplicitous.

Lawrence was very aware of duplicity in other writers when he read their works. His analysis of "classic" American literature derives its abiding strength from precisely that sort of insight. His image of the writer in *The Symbolic Meaning* is of a person who is divided, like some of Lawrence's own fictional creations, between somnambulism and wakefulness (*SM* 18). For Lawrence, the creative tension in the work of the American writers was the result of an unconscious compulsion working against the conscious intention. Much the same can be said for Lawrence's own fiction in the period between the marriage and leadership works, written roughly between 1919 and 1922. The conscious intention is legible in the metafictional comments that infiltrate his writing, as Lawrence thinks out loud, as it were. It is to revise the direction of his fiction – and this often involves literal revision of older fiction – so that it may evolve into a harsher and more robust form in keeping with his bitterly cynical new world view. The intention, however, triggers a survival anxiety that, in the end, scuppers the project and prompts Lawrence eventually – after the politically naive and tendentious *Kangaroo* and *The Plumed Serpent* – to abandon it. The "tenderness" that he tells Witter Bynner is now needed in heterosexual relations and in the fiction depicting those relations (*L* vi 321) is a long way indeed from the leadership works. How he gets from leadership to tenderness is not the story I have been attempting to tell, although it may be worth telling. What we have seen in the foregoing pages is the means by which Lawrence's fiction declined from the height of his powers in *The Rainbow* and *Women in Love*. The transitional fiction comprises part of the fulcrum that allows his fiction to teeter toward the leadership concerns and, by critical consensus, the period of his most dubious and misogynist work. In the simplest of terms, Lawrence the writer went from very good to horrid (in the words of the English nursery rhyme) because he was at one and the same time so very alive and so deathly ill. His writing violated his own interdiction against the artist allowing a "metaphysics" to skew creativity, to the extent that Lawrence became the very split consciousness he describes in *Studies in Classic American Literature*.

Only in recent years has the kind of critical distance necessary to place Lawrence in his time and place become possible. We are now, and only now, beginning to understand, for example, the full complexity of Lawrence's relationship to Bloomsbury ideas about art, as Fernihough has so ably demonstrated, to Freudian ideas about the unconscious, and, I have been arguing, to Darwinian ideas about survival. And it is through the lens of survival anxiety in general and a Darwinian perspective in particular that we may view the deleterious influence on his fiction of Lawrence's complicated relationship with women.

Notes

INTRODUCTION

1 I am aware that my rather conventional assessment of Lawrence's fictional *œuvre* is by no means a unanimous one among critics. Graham Hough, in his *The Dark Sun* (1956), attempted a rehabilitation of *Kangaroo*, and there have been many attempts since to reassess positively the novels written after *Women in Love*. Paul G. Baker, for example, urges upon us the merits of *Aaron's Rod* in a 1983 work. Peter Balbert laments the lack of critical appreciation for the "underrated" *The Lost Girl*, a "curious, powerful, and ultimately ambivalent fiction" (381). Barbara Mensch, in her *D.H. Lawrence and the Authoritarian Personality* (1991), argues that the leadership works, including *The Plumed Serpent*, are dramatizations of the kind of authoritarianism Lawrence deplores. I find a number of the arguments very interesting but none compelling enough to change my basic stance. So at the risk of engaging in a question-begging exercise, I am assuming a consensus that *The Rainbow* and *Women in Love* represent an achievement greater than the works that follow them.

2 I should make it clear that I see the transitional phase between the marriage and leadership works as comprising *The Lost Girl* (1920), the volume of three novellas published in 1923; *Aaron's Rod* (1922); and *England, My England* (1922). These are the major fictional works of the transition. The period also includes various poems and non-fictional work such as *Movements in European History* (1921) and *Fantasia of the Unconscious* (1922). The leadership period proper would then begin

with *Kangaroo* (1923) and include such works as "The Woman Who Rode Away" (1925) and *The Plumed Serpent* (1926). A salient feature of the transition is Lawrence's exploratory attempt to adapt his older vision to his new inspiration. In the leadership works themselves, there is a palpable commitment to the leadership ideal.

3 Sargent actually argues that Lawrence's work, *The Lost Girl* being the example to hand, is much less misogynist than it is normally taken to be. After decades as a post-Kate Millett foregone conclusion, the very existence of Lawrence's misogyny is now being questioned in some quarters. Although the emphasis of this study is not on revisiting the evidence for or against Lawrence as a misogynist, evidence of a basic hostility to women will, of necessity, appear along the way.

4 We do well to recall that deeply ingrained in English literary convention (because of its importance in actual life up until the early twentieth century) is the perverse dependency of upper-class individuals upon their servants for nurturing duties such as dressing, bathing, adorning, consoling, etc. that in contemporary life belong to a parent. At the same time, those mother-surrogate servants must never forget their "place" beneath their masters and mistresses in the class hierarchy of English society. Such an ingrained ambiguity in familial roles, so pervasive in the English novel, makes it difficult to separate Lawrence's exploration of the mother-child hierarchy and his imaginative inversion of the power distribution within it from his numerous attacks on the English class system. However, Lawrence's own psychopathology tilts his efforts ultimately in the direction of misogyny rather than social reconstruction.

5 One example of poststructuralist criticism is a collection of essays in the Longman Critical Readers series edited by Peter Widdowson called simply *D.H. Lawrence* (1992). Widdowson's introduction is entitled "Post-modernising D.H. Lawrence." Another example of a critic using such a perspective is Joyce Wexler's "D.H. Lawrence Through a Postmodernist Lens" (1998). As for the latter, we may turn, for example, to Howard J. Booth's "'Give me *differences*': Lawrence, Psychoanalysis, and Race" for a study that touches upon the postcolonial questions raised in Lawrence's work.

6 Emile Delavenay points out in a 1987 essay that Russell's remark "was not intended to accuse Lawrence of complicity with the Nazis" but rather to condemn the "Germanic modes of thought" that were adopted by Lawrence and *were* responsible, ultimately, for Auschwitz ("Further Thoughts" 174). However, Mark Kinkead-Weekes's biography of Lawrence's middle years places any scruples Russell may have had when he made this remark and, indeed, his motives entirely in doubt (810n68).

7 Wayne Templeton estimates that Lawrence "extended by about ten years the average life expectancy of a tuberculous victim before the advent of antibiotics" through sheer force of will (194).
8 "Getting On" is an unpublished, early version of Lawrence's essay "Return to Bestwood," probably written early in 1927.
9 Although the practice may be a little confusing, I refer throughout this study to the volume of novellas as the "*Ladybird* novellas," but the parenthetical reference to the volume will be "*Fox*." The first British edition of the book was entitled *The Ladybird*; the first American edition was called *The Captain's Doll*. For reasons they do not discuss, the editors of the Cambridge University Press edition of the volume, which I have used as my text, have chosen to begin the book with *The Fox*, and my parenthetical references reflect the resulting title of the book.

CHAPTER ONE

1 Levine is paraphrasing from Sally Shuttleworth's book *George Eliot and Nineteenth-Century Science*, Cambridge: Cambridge University Press, 1984.
2 It was in December 1907 that Lawrence's first publication, the story "A Prelude," appeared in the *Nottinghamshire Guardian* under Jessie Chambers' name.
3 Other remarks by Lawrence on evolution in fiction, essays, letters, and poems abound. The following sampling is neither systematic nor exhaustive, but it does suggest a continuing and developing engagement on Lawrence's part with evolutionary theory. In his "Study of Thomas Hardy" (written in 1915, published in 1936), Lawrence discusses and rejects the concept of evolution promulgated by the great scientists or thinkers of the preceding generation, Darwin, Spencer, and Huxley (*P* 485). At the same time, his suggestion of a movement from undifferentiation toward greater and greater differentiation of life (*P* 432) is clearly similar to Spencer's description of the process of evolution. In the posthumously published *Mr Noon* (written in 1920, published initially in 1934 and then in restored form in 1984), Lawrence describes how his alter ego, Gilbert Noon, wonders at the beauty of a tree: "What he loved so much in plant morphology was, that given a fixed mathematical basis, the final evolution was so incalculable. It pleased him to trace inherent individual qualities in each separate organic growth, qualities which were over and above the fixed qualities belonging to the genus and the species, and which could not really be derived by a chain of evolutionary cause-and-effect. Could they?" (*MN* 43–4). In *Fantasia of the Unconscious*, Lawrence remarks that he does not believe in evolution but rather in the "rainbow-change of ever-renewed creative civilizations" (*FU* 14); however, later in the

same essay he suggests that new growth requires the death of the old (*FU* 180). "Him with His Tail in His Mouth" (1925) presents a light-hearted view of evolution that includes the poem that later appeared in the February 1926 *Adelphi* under the title "Creative Evolution" (PII 428–31). In contrast, the epilogue to *Movements in European History* (written in 1924, first published in 1971) contains a much more sweeping and authoritative statement: "We can no longer believe that man has descended from monkeys. Man has descended from man" (*MEH* 255). In the essay "On Human Destiny" (1924), there is a rather odd conflation of the ideas of a static Great Chain of Being and devolutionary movement: "Man is a domesticated animal that must think. His thinking makes him a little lower than the angels. And his domestication makes him, at times, a little lower than the monkey" (*PII* 623). In the poems "Cypresses" from *Birds, Beasts and Flowers* (1923) and "Dies Irae" and "Nullus" from *Pansies* (1929), there are statements on evolution or its various doctrines, not to mention the entire implicit argument of the former volume, which has been seen as evolutionary in foundation (Heywood 88, 90, 102). In a late essay, "Introduction to These Paintings" (1929), Lawrence unaccountably suggests that his apostasy came "[a]fter years of acceptance of the 'laws' of evolution" (*P* 575).

4 We may compare Anne Fernihough on this point: "[Lawrence] connects chronological time with oppression and the domination of nature" (171).

5 Lawrence tended at times to equate his own illness and that of his culture. In a letter to Cynthia Asquith of 5 August 1915, he writes: "I was so sick I felt I could not come back: all these little, amorphous houses like an eruption, a disease on the clean earth … The whole thing like an active disease, fighting out the health … One feels a sort of madness come over one, as if the world had become hell. But it is only superimposed: it is only a temporary disease. It can be cleaned away" (*L* ii 375, 376).

6 The term "allotropic" was used by Lawrence in his famous June 1914 letter to Edward Garnett, in which he sets out his experimental sense of character in the book that became *The Rainbow* in response to Garnett's dismay at Lawrence's obvious disinterest in creating fictional characters like the ones he had been so successful with in *Sons and Lovers* (*L* ii 183).

7 At times, Lawrence was capable of quite astonishing statements regarding the necessity to kill. In "Return to Bestwood" (first published in 1968), for instance, he suggests that "[h]opeless life should be put to sleep, the idiots and the hopeless sick and the true criminal" (*PII* 265). In an early letter (from 1908), moreover, a rather enthusiastic support for eugenics is discernible: "If I had my way, I would build a lethal chamber as big as the Crystal Palace, with a military band playing softly, and a Cinematograph working brightly; then I'd go out in the

back streets and the main streets and bring them in, all the sick, the halt, and the maimed" (*L* i 81).

8 Mensch is referring here to an idea in David Boadella's *The Spiral Flame: A Study in the Meaning of D.H. Lawrence* (Nottingham: Ritter Press, 1956). Boadella, in turn, is using a concept from Wilhelm Reich.

9 We may compare the following statements on class: first, from "Autobiographical Sketch": "Class makes a gulf, across which all the best human flow is lost" (*PII* 595); second, from Lawrence's review of Trigant Burrow's *The Social Basis of Consciousness*: "And the last great insanity of all, which is going to tear our civilization to pieces, the insanity of class hatred, is almost entirely a 'normal' thing, and a 'social' thing" (*P* 381).

10 Fernihough finds the source of the rapid fall in Lawrence's reputation in the "recognition of misogynistic elements" in his work and in "the flourishing and demise of a post-Romantic critical idiom deeply informed by notions of the organic" (5).

11 This connection seemingly undermines Ebbatson's contention that "Lawrence's work centres upon the clash between the rationalist-irrationalist reading of the Universe expounded by Darwinism and the transcendental-vitalist reading of the Romantic Nature tradition" by suggesting that, in fact, the former supports the latter (*Nature Tradition* 258).

12 In many of the transitional works, there are two or three generations whose contrast represents the telescoped evolution that Lawrence wishes to impose on his characters. Where a third generation is present, the children are usually ciphers with occasionally significant symbolic value, as in "England, My England," where the crippling of the eldest daughter is an emblem of the *inherited* weakness passed on by Egbert to his daughter and actualized dramatically for the father in his suicidal death in the trenches. The members of the older generation usually belong to one of two sorts. There are the vital, old rapscallion men – like Lord Beveridge of *The Ladybird* – who have some of the features of the vital young men at the centre of the story but who have been repressed or have seen time pass them by. Then there are the vitiated old people who, like Sir William Franks in *Aaron's Rod* or Miss Frost in *The Lost Girl*, must be weeded out in order to dramatize the necessity for a new sort of being if the individual and species are to survive.

13 By 1921, Judith Ruderman points out, "animals become the controlling symbols in Lawrence's fiction, and accordingly he then gave his works animal titles" (184).

14 In "Merging – with Fish and Others," David Cavitch speaks of Lawrence's "increasingly sadistic orientation to external things" (177).

15 The distinction between power and force is most elaborately set out in "Blessed Are the Powerful" (1925), where Lawrence writes, among much else, that power is not "issuing loud orders and getting your own way, doing your opponent down" (*PII* 439). The difference is also discussed briefly in the epilogue to *Movements in European History* (*MEH* 263).

16 I have argued elsewhere ("'Jews of the Wrong Sort': D.H. Lawrence and Race") that Lawrence's racism must be seen in the context of his misogyny and as an expression of feelings of vulnerability with regard to ego boundaries. I discuss this issue briefly in chapter 3.

17 It is arguable, however, that Lawrence's concept of blood knowledge is disguised racism and classism. After all, how different from the ideas of "blue" blood or racial blood lines is blood knowledge? We can see the most extreme statement of this view of Lawrence in Jascha Kessler's essay: "No wonder Lawrence's later books were popular in Nazi Germany: his blood theory led him directly into totalitarian ideology. For to Lawrence blood was not merely a trope, or a spiritual symbol: it was the quintessence of the racial" (484). I think we can fashion a corrective to this still fairly widely held view of Lawrence the Fascist by demonstrating that Lawrence's frequently unthinking comments on race were part of a larger and more serious exploration of hierarchy that in the last instance emphasized individual abilities rather than racial differences.

18 In his wedding speech, Tom says: "'*If* we've got to be Angels, ... and if there is no such thing as a man nor a woman amongst them, then it seems to me as a married couple makes *one* Angel ... For ... an Angel can't be *less* than a human being. And if it was only the soul of a man *minus* the man, then it would be *less* than a human being'" (*R* 129).

19 In his "Introduction to These Paintings," Lawrence writes that "Cézanne's apples are a real attempt to let the apple exist in its own separate entity, without transfusing it with personal emotion. Cézanne's great effort was, as it were, to shove the apple away from him, and let it live of itself. It seems a small thing to do: yet it is the first real sign that man has made for several thousands of years that he is willing to admit that matter *actually* exists" (*P* 567–8).

20 This idea was first brought to my attention in Lisbeth Arndal's unpublished dissertation, pages 38, 66.

21 As Wayne Templeton notes, one way of contracting tuberculosis, especially in infancy, is through ingesting contaminated milk (190).

22 In his study of the effect of illness on Lawrence's writing, Templeton describes Lawrence's "coping mechanism" as one that allowed him "to focus on a quest for strength and sources of strength other than the usual or the obvious" (193).

23 We may compare an observation of Philip Heseltine (who was friendly with Lawrence for a brief time during the war and later denounced him) that Lawrence tended to treat people as though they were characters in one of his novels (Nehls i 352).

24 The moon is not explicitly the cause of Mrs Hepburn's fall, but when the captain explains to Hannele how the accident happened, he says that his wife "'could never bear even to look out of a high window. Turned her ill instantly if she saw a space below her. She used to say she couldn't really look at the moon, it made her feel as if she would fall down a dreadful height'" (*Fox* 110).

25 See particularly Paul Delany's *D.H. Lawrence's Nightmare* (1978).

26 See, for example, Jeffrey Meyers' useful appendix "A History of Illness" under the entries for the war years (*D.H. Lawrence: A Biography* 389).

27 Fairly late in life, Lawrence wrote in a letter that "[i]t's time there was an *enormous* revolution – not to instal soviets, but to give life itself a chance" (*L* vii 99).

28 Daniel J. Schneider argues for the influence of Houston Chamberlain on Lawrence's thinking about evolution in connection with their similar views on the individual ("Houston Chamberlain" 166). The connection between Lawrence and Chamberlain was earlier established by Emile Delavenay in his *D.H. Lawrence: The Man and His Work* (English translation 1972), as Schneider, of course, realizes and acknowledges. The thrust of his note is to suggest that Delavenay distorts the influence in a number of important areas. The discussion of evolution is cursory.

CHAPTER TWO

1 In a letter dated 7 December 1921 to his agent Curtis Brown, Lawrence discusses the three novellas (*The Ladybird* was still being written at this point) and specifies that "[t]hese three I want to go in one volume by themselves" (*L* iv 143). To Robert Mountsier a few weeks later, Lawrence writes that he has finished the last of the novellas and that with the three of them, "I think you have a rather nice volume" (*L* iv 150).

2 Graham Holderness has presented one of the novellas, *The Fox*, as part of a Lawrentian transition. It is, however, a much different transition than the one I am dealing with. In fact, as I briefly mentioned in the last chapter, Holderness ignores the leadership works. Very briefly stated, he groups *The Fox* together with the "Study of Thomas Hardy" and *The Rainbow* as the major texts taking Lawrence from realistic tragedy to symbolic myth. While *Women in Love* returns Lawrence to the realistic mode, *Lady Chatterley's Lover* is yet another foray into the myth of an organic community annunciated in *The Rainbow*. While I find much to admire in Holderness's attempt at a materialist reading of a

number of major Lawrence texts, I believe that his characterization of the mythic qualities of the Marsh Farm glosses too lightly over the dramatization of real crises, for example in the marriage of Tom and Lydia. *The Rainbow* is a difficult text partly because of its blending of the realistic with the mytho-symbolic. Furthermore, Holderness's criticism of Lawrence's escape from painful social realities into a myth of an organic community wholly contained within the individual does an injustice to Lawrence's honest if ill-fated attempts to imagine a revolution in the very concept of class.

3 *The Captain's Doll* is a longer and much altered version of "The Mortal Coil" (1917); *The Fox* is a longer version of a short story by the same name published in November 1920, and also owes something to "Hadrian" (also known as "You Touched Me") published in April 1920; and *The Ladybird* is a revision and expansion of "The Thimble" (1917). Elgin W. Mellown's article on the origins of and allusions in *The Captain's Doll* briefly compares "The Mortal Coil" with the novella that developed out of it, Patrizio Rossi has compared the texts of the short and long versions of *The Fox*, and James F. Scott has done something similar with *The Ladybird*. In addition, Laurence Steven compares "The Thimble" and *The Ladybird* in a study of the short story. In the autumn of 1921, barely two years after finally receiving permission to leave an England he had long been desirous of fleeing, Lawrence refashioned these works in a retrospective mood. In his letters of the time, he repeatedly speaks of getting his manuscripts in order so that he might be free to set out anew, away from England. See, for example, the letter to S.S. Koteliansky of 10 November 1921 (*L* iv 113–14) and the letter to E.H. Brewster of 16 November 1921 (*L* iv 124–6).

4 As Monroe Engel has pointed out, there is an association in the novellas between the old form of love and the concepts of sanity and over-responsibility (94). What we should notice is that while a positive portrayal of madness or irrationality has been part and parcel of Lawrence's championing of the life of the body from the very beginning of his writing career, the denunciation of female responsibility is new and characteristic of this transitional period and of the leadership works that follow it. It is related in the three novellas to Lawrence's attempts to invert the parent-child hierarchy of power.

5 In an article published in 1972, Gurko suggests that the linking of the three novellas in one volume challenges the reader to find the "shared principle" that holds them together (174). He mentions the wartime, the European setting, and the folkloric/mythical background suggested by other critics: the Sleeping Beauty motif that Harry T. Moore (in *The Intelligent Heart*) identified in *The Ladybird* and the Pluto-Persephone myth advanced by George Ford (in *Double Measure*). One might argue

that the association between woman and sleep/darkness in these tales has as much to do with the emergent leadership theme as with folklore or mythology. As Daniel J. Schneider has suggested, "[n]ight, for both Nietzsche and Lawrence, is associated with the female and with love; day with the male's creative activity in the building of a new world" (*Artist as Psychologist* 47). Gurko himself suggests that the common central problem is that the three male protagonists must determine "how to escape from their imprisonment as caricatures and regain their living wholeness" (176). The caricatures are emblematized in the titles. Two further common features holding the volume together are the death-resurrection motif and the endings of imminent departure that suggest a geographical analogue to the male-female conflict in the tales.

In "The Continuity of Lawrence's Short Novels," an essay published in 1963 and encompassing *St Mawr* and *The Man Who Died* as well as the three *Ladybird* novellas, Monroe Engel suggests that Lawrence is writing "thesis fiction" in his short novels. In each, he claims, one may find the following elements: "the objective and fluent style; analogy – generally animal analogy; disordered relationships; the opposition of traditional love and a new kind of relationship between the sexes ...; the use of magic; and the visionary ending associated with the emergent theme of resurrection" (99–100).

6 Paul Delany quotes from a letter Lawrence wrote to J.M. Murry in January 1916, in which Lawrence's identification with a hunted fox is clear: "I feel now pushed to the brink of existence, and there remains only to fall off into oblivion, or to give in ... I feel absolutely run to earth, like a fox they have chased till it can't go any further, and doesn't know what to do" (*Nightmare* 188).

7 There are many other examples of the presence of food, literal or metaphorical, in the novella, including the following: when Henry sets off for his fifty- or sixty-mile bicycle ride back to Bailey Farm after receiving March's letter, it is "without a thought of food" (*Fox* 60); when he is about to fell the dead tree, a number of ducks run into a dangerous area and Banford shoos them away by saying "'There's no food. There's nothing here. You must wait a bit'" (*Fox* 64).

8 Critics have made much, perhaps too much, of the latent lesbian relationship between March and Banford, but certainly it appears that Lawrence intends Henry to offer March a more "natural," phallic option to Banford's love. See Patricia C. Davis, "Chicken Queen's Delight: D.H. Lawrence's 'The Fox'" for an interesting interpretation of the alignment.

9 In the first version of the story, the two women are said to be "over thirty" ("The Fox" 28), whereas in the published novella, they are "near thirty" (*Fox* 7). In both short story and novella, Lawrence writes that

Henry "would not be more than twenty" ("The Fox" 35; *Fox* 14), although in the earlier version, in response to March's objection that she is old enough to be Henry's mother, Henry says "'I know how old you are ... You're thirty-three – and I'm nearly twenty-one. That's not old enough to be my mother'" ("The Fox" 46). On the last page of the short story, March feels that Henry is "so young – and so old – so old" ("The Fox" 47).

10 Darwin points out that although "[v]ariability is not actually caused by man, ... [i]t is certain that he can largely influence the character of a breed by selecting, in each successive generation, individual differences so slight as to be inappreciable except by an educated eye. This unconscious process of selection has been the great agency in the formation of the most distinct and useful domestic breeds. That many breeds produced by man have to a large extent the character of natural species, is shown by the inextricable doubts whether many of them are varieties or aboriginally distinct species" (432).

11 We may compare the scene with the dead leaves in *The Fox* with the following description of the renewal of arboreal life from *Fantasia*: "But each new tip arises out of the apparent death of the old, the preceding one. Old leaves have got to fall, old forms must die" (*FU* 180).

12 In "Blessed Are the Powerful," Lawrence wrote: "If you're a fox, like the born financier, you get a bellyful and more. If you're a stork, or a flamingo, or even a *man*, you have the food gobbled up from under your nose, and you go comparatively empty" (*PII* 441). While the fox figure in this passage has associations not found in *The Fox*, the connections to survival and food are nevertheless striking.

13 H.M. Daleski writes of Henry killing Jill Banford: "We can only conclude, I think, that when Lawrence, who has such a reverence for life, can be taken to justify murder, it is because the murder is incidental to a compulsive justification of something else" ("Aphrodite" 154). Daleski here is guilty of a readily understandable begging of the question, for it is not clear that Lawrence in all moods and at all times did reverence all manifestations of life. If one looks at the changes made to the original version of *The Fox* in 1921, one sees that Lawrence has made Banford less "fit" to survive in the novella. Nevertheless, Daleski's phrase "compulsive justification of something else" is useful. My contention is that the "compulsion" is the parent-child relationship, the "justification" a world badly askew, and the "something else" a redefined hierarchy of existence.

14 See my article "A Second Caveat: D.H. Lawrence's *The Fox*" for a full exposition of this connection.

15 March and the text itself both clearly associate the fox with Henry, and most commentators on the story accept that connection. However, Jan

Good suggests that the animal represents both "the masculine element in Henry" and "the masculine phallus within March's psyche which must be killed" (220). For the reasons that I set out, I believe that the most significant representational value of the fox is connected to the idea of survival.

16 Mansfield's comment was actually directed at *The Lost Girl*.

17 Lawrence describes Tess in his Hardy study as "out of place, utterly out of her element and her times" (*P* 486).

18 Eugene W. Dawson views the entire scene as a compelling explanation for Hepburn's original attraction to his wife (144). Gerald Doherty recognizes the problem, but he tries to account for what he terms Hepburn's "cryptic 'biography' of his wife" by arguing, not very convincingly in my view, that the scene indicates Hepburn's ability as "a prolific metaphor-maker" ("Figural Play" 9).

19 In the ancestral tale to *The Captain's Doll*, "The Mortal Coil," Lawrence several times describes the protagonist, the German officer Friedeburg, in terms of imagery of confinement and release: "Sweet it was to pass through the gates of the town, the scanty, loose suburb, into the open darkness and space of the country" (*PII* 80). However, the heroine, Marta Hohenest, is also described in the same terms: "She felt terribly bound, walled up" (*PII* 67). The difference seems to be that whereas Marta can be "released free into the infinite of ecstasy" by sex with Friedeburg (*PII* 75), he himself requires the purposive life with his soldiers. While purposive life in this early story (the first version was written in 1913) is portrayed as destructive of sexual fulfillment albeit attractive, in *The Captain's Doll*, male purpose is a necessary precondition to a full sexual life.

20 In a letter to Marie Hübrecht written on 8 November 1920, for example, Lawrence speaks of his mother-in-law and all of Germany being "threadbare" (*L* iii 620). In August 1921, he writes to Catherine Carswell from the Zell-am-See, the setting for the climax of *The Captain's Doll*, and describes Germany's wartime ally Austria as "financially and commercially just ruined" (*L* iv 64). To his mother-in-law in mid November 1921, Lawrence writes of his intention to send money and tea, and, in early December of the same year, he expresses satisfaction that she has received his cheque (*L* iv 121, 122, 147).

21 Hepburn's given name is surely carefully chosen by Lawrence. In "Democracy," he speaks of two ways in which the human passion for including everything within the self can be realized: "The first is Alexander's way, the way of power, power over the material universe" (*P* 707). Elgin W. Mellown has much to say regarding the names of the two principal characters in *The Captain's Doll*. He relates the name Alexander to the obvious idea of conquest and also suggests that

Lawrence likely knew "the familiar story told by some medieval writers which held that Alexander was susceptible to the wiles of women" (232). In such a case, I feel it is safer to refer to what Lawrence himself wrote elsewhere regarding the name.

22 Ascent of another sort may be associated with Hannele if we accept as a source for Lawrence's novella Hauptmann's *Hanneles Himmelfart* (1892, tr. 1914), literally "Hannele's Flight to Heaven," as Mellown suggests (230–2). Hepburn mentions the play (*Fox* 96).

23 The literal and symbolic meaning of the doll in this novella has been debated for a long time. Several commentators have related the doll to Ibsen's *A Doll's House*, among other works (Mellown 228, McDowell 144), while others have seen it variously as a kind of voodoo prop (Widmer 158) or the embodiment of love as adoration (Leavis, *D.H. Lawrence* 210) to give two examples. Although the meaning of the doll is multiple, perhaps the best single gloss on it comes from Lawrence himself, who equates the concept of a mannikin with the excesses of egoism in "… Love Was Once a Little Boy" (*PII* 458).

24 See, for example, Robert H. MacDonald's article (144), and compare Doherty ("Figural Play" 13), Eugene W. Dawson (147), and Frederick P.W. McDowell (154). H.M. Daleski relates the imagery of the "insatiable" water to Lawrence's rejection throughout the volume of novellas of an active female sexuality imaged by "Aphrodite of the foam" in *The Ladybird* ("Aphrodite" 157). While in itself this is a useful observation, it is also another good example of a critic ignoring the literal for the metaphorical meaning of words like "insatiable." Such a focus leads Daleski to view *The Captain's Doll* as unmarred by the flaws that vitiate the two other stories ("Aphrodite" 158).

25 Gerald's infantile dependence upon Gudrun shortly after the death of his father is starkly described: "Mother and substance of all life she was. And he, child and man, received of her and was made whole … Like a child at the breast, he cleaved intensely to her, and she could not put him away" (*WL* 344–5).

26 Herbert Howarth is apparently taken in by Lawrence's fudging. In a comparison of *The Trespasser* and *The Captain's Doll*, he argues that Siegfried is a child-man while Hepburn is an adult (224–5). To be fair to Howarth, it should be added that he does suggest that Hepburn's victory is "rather less secure than the story-line claims" (225).

27 Daphne's family background certainly sounds like that of the Lawrences in Eastwood: "[H]er whole will was fixed in her adoption of her mother's creed, and in condemnation of her handsome, proud, brutal father who had made so much misery in the family" (*Fox* 161). Although Lady Daphne is modelled largely on Lady Cynthia Asquith, the fictional character's weak lungs further emphasize the association

between character and creator. Under Dionys's "spell" but before any sexual consummation, Daphne feels that the habitual pressure against her breathing has been removed (*Fox* 213).

28 Ruderman suggests that Daphne deliberately makes Dionys feel small and insignificant: "this is the trick of the 'devouring mother,' who thwarts her child's attempts at independence" (80).

29 The Satanic associations of both Daphne's father and her eventual lover suggest a classical oedipal father-daughter connection. As in the Hepburn-Hannele relationship, there are a good many years between Dionys and Daphne. Nevertheless, the oedipal concern does not seem to interfere with the pre-oedipal theme of nurturing through nursing.

30 The oral element is not entirely lacking in *The Ladybird*. Lawrence Jones relates a passage in *Fantasia* to Lawrence's description of Count Dionys's teeth, which seem "white and powerful" to Daphne (*Fox* 185). In *Fantasia*, Lawrence laments the loss in modern humanity of "the sharp and vivid teeth of the wolf, keen to defend and devour" (*FU* 61). See Jones (4–5).

31 The Christian-pagan opposition I am suggesting in no way undermines Kingsley Widmer's argument in *The Art of Perversity* that Lawrence is opposing a Nietzschean Dionysian principle to an Apollonian one in the person of Basil (whom Widmer calls "Boris") (47). Compare chapter 2 of James C. Cowan's *D.H. Lawrence and the Trembling Balance* (1990). The two oppositions are developed simultaneously, but only the Christian-pagan one is relevant to my argument regarding hierarchical inversions. The hammer may have Nietzschean overtones as well, since the alternate title to *Twilight of the Idols* is "How One Philosophizes with The Hammer."

32 Basil's character is based loosely upon the son of the prime minister during the early part of the war, Herbert Henry Asquith. Asquith was made first earl of Oxford and Asquith, but can hardly be said to have belonged to the hereditary aristocracy.

33 Zipes is summarizing an article by Peter Taylor and Hermann Rebel entitled "Hessian Peasant Women, Their Families, and the Draft" published in *Journal of Family History* 6 (1981): 347–78. Zipes finds their argument interesting but flawed (49).

CHAPTER THREE

1 Clark writes: "Lawrence saw himself in direct succession from the classic Americans, rejecting the modernist experiments of his contemporaries and putting behind him the only tradition he had ever recognized: the English novel from George Eliot to Thomas Hardy. His own discoveries in how actions spring from subliminal sources he could take

with him to the form he now adopted and updated: the 'romance' of classic American fiction. During eight years of experiment with this form in the pilgrimage novels, Lawrence's essays on his American predecessors went through revision after revision" (195).

2 Daleski is referring specifically to *Aaron's Rod*, and he continues: "for he is engaged in asserting and defending a 'male' doctrine of 'power,' which in its kindred manifestation as a creed of 'love' he violently condemns, and which, anyway, runs counter to his deepest 'feminine' inclinations to 'be'" (*Forked Flame* 210).

3 This is from the original version of "The Spirit of Place," which appeared in the *English Review* of late 1918 and was later revised and reduced when it became part of *Studies in Classic American Literature*. In this version, Lawrence distinguishes between two meanings of "art-speech" in American literature: the manifestation of the conscious intention of the writer and the symbolic product of the unconscious.

4 For example, G.M. Hyde details the engagement of Lawrence's novel with Bennett's in what amounts to a Bloomian "strong reading" of the earlier novel by Lawrence (80). Sargent suggests that Alvina Houghton of Woodhouse is Lawrence's rewriting of Jane Austen's Emma Woodhouse (189–90).

5 Mark Kinkead-Weekes suggests that "[t]he story behind the story was the Insurrection of Mr. Lawrence" (574), this fictive title alluding to the working title of *The Lost Girl*, "The Insurrection of Miss Houghton."

6 G.M. Hyde links the narratorial voice in *The Lost Girl* to the music-hall tradition: "At every stage, Lawrence's text draws attention to its own coming into being, its 'mode of production,' by means of a gesturing, self-dramatising narrator who, like a music-hall entertainer, enacts the disconcerting switches of code that frustrate and disorientate passive readerly expectations" (81).

7 Most commentators have viewed Mr May as a portrait of Lawrence's acquaintance Maurice Magnus, but David Holbrook suggests that he may be "a kind of caricature of Lawrence himself, sexually" (275).

8 One contemporary review, in *The Westminster Gazette*, suggested somewhat waggishly that "Mr. Lawrence's muse must have mated with Mr. Arnold Bennett or with Compton Mackenzie, or with both" (qtd in G.M. Hyde, 76).

9 Gary A. Wiener argues for a detailed connection between *The Lost Girl* and Blake's "The Little Girl Lost" and "The Little Girl Found" in *Songs of Innocence and Experience*. He suggests, for instance, that Alvina's adopted name, Allaye (often viewed as a sexually suggestive pun), may derive from the latter poem's description of Lyca's parents meeting a lion: "But their fears allay, / When he licks their hands; / And silent by them stands" (247). What is more significant, in the context

of *The Ladybird*'s Dionys Psanek wishing for a new name and of the later revolutionary *noms de guerre* in *The Plumed Serpent*, is the change in name itself rather than what name is adopted.

10 Judith Ruderman claims that "a most important cause of Alvina's troubles" is "her lack of a father … Alvina's father abdicates his leadership responsibilities and, in effect, abandons his daughter from early babyhood" (39). Ruderman views Alvina's relationship with Dr Mitchell as an inversion of the mother-child relation (44), as it potentially is. But I would argue that Alvina is attracted to Dr Mitchell initially because he is an authority figure and thus a plausible father substitute as much as because of his boyishness. In the end, it is the prospect of his dependency that frightens her off as well as the fact that the disjunction between his appearance of power and authority and the reality of his weakness remind Alvina of the keeping-up-appearances mentality that she is fleeing from. Alvina has pretty well made up her mind not to marry Dr Mitchell by the time he reveals his infantilism upon her refusal to put on his engagement ring.

11 We may compare this dynamic to two of the novellas. Here Alvina seems attracted *because of* an age difference between her and the good doctor. In *The Fox*, Henry goes to great lengths to *deny* the importance of the age difference between himself and March (March being about ten years older). In *The Captain's Doll*, Hepburn flees from a mother-wife figure who is older than he to one who is considerably younger. That the protagonist in *The Lost Girl* is female complicates the comparison, but may mirror societal pressures on women of "breeding" age. Holbrook is puzzled by the fact that Alvina is sensible enough to reject Dr Mitchell when he turns violent but is prepared to accept the violence in Ciccio that manifests itself in his murderous attack on Max, another of the Natcha troupe, as well as in his sexual domination of Alvina (Holbrook 273*f*). What Holbrook fails to take into account is that Alvina is horrified less by Dr Mitchell's violence than by the infantile dependency that he reveals once his mask of reserve has slipped off. As Ruderman suggests, Alvina's affair with Dr Mitchell "seems to invert the pattern of mother-child relationships established in the novel: no longer is Alvina the child dependent upon a mother whom she seeks to please; instead, she herself becomes the divine mother adored by a child-slave" (44).

12 See Virginia Hyde's "'Lost' Girls: D.H. Lawrence's Versions of Persephone" for a full exposition of Lawrence's use of the Persephone figure in his writing, including *The Lost Girl*.

13 It is not Darwin's view that the universe, or more particularly nature, is a machine. As George Levine has shown, Darwin broke with the natural theologists of his time essentially on this point. Unlike them,

"his model is the tree, not the machine ... Darwin, though an inheritor of eighteenth-century materialist thought, had nevertheless absorbed the organicist assumptions of the romantic poets" (Levine 50). Lawrence, whose model is also the tree, may, however, have viewed Darwin's theories as overly mechanistic. It is also possible that he is satirizing Mrs Tuke's knowledge of science.

14 Balbert suggests that this "scene of Lawrencian communion" is similar to "analogous moments of benedictive, renewing touch" in a number of other works of fiction by Lawrence (389), but that here the opportunity at real communion (they are in a church, after all) is squandered and "Alvina is, in effect, prostituted by a man who returns nothing for her care of his flesh" (390). There is also the possibility that Lawrence here is self-reflexively and rather bitterly commenting on his own earlier depictions of such "benedictive" moments.

15 Jeffrey Meyers compares her to Proust's Tante Léonie (*Experience of Italy* 96).

16 Virginia Hyde maintains that Alvina "has by no means been raped" ("Versions of Persephone" 110). I agree that the act is much more ambiguously presented than the rape in Lawrence's "The Princess," for example, but Ciccio certainly ignores Alvina's pleas to desist, and we are left in little doubt that it is Ciccio's superior physical strength that determines the issue: "For a second, she struggled frenziedly. But almost instantly she recognized how much stronger he was, and she was still, mute and motionless with anger" (*LG* 233).

17 *The Ladybird* begins: "How many swords had Lady Beveridge in her pierced heart!" But she has determined that her heart "should never die" until, when power slips from her in the last years of the war, "it seemed as if the many swords had gone home into the heart of this little, unyielding Mater Dolorosa" (*Fox* 157).

18 Hilary Simpson characterizes Alvina's extinguishment in her love for Ciccio as, in effect, the culmination of a process begun in *The Fox*: "We see realised in her [Alvina] the state that Henry had demanded of March in 'The Fox'" (77).

19 Holbrook, for example, deems it likely that Ciccio will never come back from the war and that Alvina will die in childbirth (276), while Ruderman suggests that the novel ends with a "promise for a future life in which the proper balance in human relationships will be achieved" (47).

20 Bonds looks at the relationship between the portrayal of self and Lawrence's own ideas about and multivalent use of language with reference to *Sons and Lovers*, *The Rainbow*, and *Women in Love*. Fernihough's exploration of Lawrence's deconstructive language use is done within the context of a larger study of Lawrence's commentary on aesthetic form in such key works as *The Symbolic Meaning*, *Studies in Classic*

American Literature, and *Fantasia of the Unconscious* and how those aesthetic principles pit Lawrence against Freud, Bloomsbury, and Martin Heidegger. The studies of Bonds and Fernihough are complementary.

21 Many details in the novel are similar to particulars in all three novellas, as Lawrence's mind circled around the same cluster of concepts during the transitional period. In terms of *The Ladybird*, first of all, Jim Bricknell's father in *Aaron's Rod* physically resembles Count Dionys in that "every strand stood distinct" in "his sparse but strong-haired white beard" (*AR* 27). Jim Bricknell's attraction to Christ and to love and sacrifice, at least the lip-service he pays to them, is similar to the devotion of Lady Beveridge and later that of Basil. The "call-note" of Aaron's flute that so attracts the Marchesa is very similar to Dionys's singing that brings Lady Daphne to his room in *The Ladybird*. As for *The Captain's Doll*, Captain Herbertson might be likened to a shell-shocked Captain Hepburn. Lilly's declaration that "'[e]verybody ought to stand by themselves, in the first place … They can come together, in the second place, if they like'" (*AR* 91), the ostensible theme of *Aaron's Rod*, is remarkably similar, despite a slight change in emphasis, to the narrator's comments in *The Captain's Doll* that "[w]e must all be *able* to be alone, otherwise we are just victims. But when we *are* able to be alone, then we realise that the only thing to do is to start a new relationship" (*Fox* 115). Aaron, like Hepburn, feels "the bonds and ligatures which bound him to the life that had formed him, the people he had loved or liked" breaking (*AR* 178). Finally, Julia in *Aaron's Rod* resembles March in *The Fox* in her "dazed" look and puckering (of the eyes rather than the mouth) (*AR* 28). Lottie, Aaron's wife, complains that Aaron, much like the Henry Grenfel Banford sees, is "selfish through and through" (*AR* 43). In fact, exactly the same words are used in *The Fox* (*Fox* 38). Lilly is even likened at one point to "a fox slipping alert among unsuspecting cattle" (*AR* 111). Julia is five years older than Robert, her estranged husband. Thus the marriage of an older woman to a younger man, the pattern we saw in *The Fox*, in *The Captain's Doll*, and in *The Lost Girl*, has been reduced to a very minor subplot in *Aaron's Rod*. I am not speaking here of a literal progression from the novellas to *Aaron's Rod*. In fact, the revisions to both occurred at about the same time, although one version of the novel was completed before the novellas were done. The point I am making has to do with the somewhat obsessive replaying of ideas and even minor details in a number of the works of the period.

22 In a recent issue of the *D.H. Lawrence Review*, John Turner reprints James White's account of the "punch in the wind" episode, which took place when he visited the Lawrences in the spring of 1918. Turner found the account in an article White published in A.R. Orage's *The New English*

Weekly of 14 July 1932. Turner also discusses White's autobiography, *Misfit* (1930). It is an interesting piece, but Turner, to my mind, fails to discuss White/Bricknell within the overall structure of the work, of which Jim's punch in the wind and his spinelessness are a part.

23 In his 1965 book, *The Forked Flame*, H.M. Daleski suggested that *Aaron's Rod* led one to the conclusion "that Lawrence is now no longer concerned with form in the novel" (211).

24 Compare Peter Scheckner: "The novel is self-reflective: on occasion the main character asks, almost of the writer himself: You have stripped me of everything; now what do I have left and for what should I live?" (110).

25 In *Women in Love*, for example, Birkin's stoning of the moon in "Moony" is the signature. The concentric circles that emanate from the image of the moon on the water, I would argue, describe the hidden structure of the fiction. For Birkin, the moon on the water may suggest what he considers to be Ursula's overdeveloped will, but as a self-reflexive signature it posits the interdependence of the fates of the individual, couples, society, particular races, and humanity as a species. So, for example, Gerald's own fate and personality impinge upon the mutually destructive relationship with Gudrun, with his social role as industrial magnate, with his symbolic fate as a representative of the "northern" white races, and, together with other characters in "Breadalby," the possible extinction of humanity as a species. The interdependence is developed symbolically primarily in chapters where water is important, apart from "Breadalby" and "Moony," in "Diver," "Sketch Book," "An Island," "Water Party," "Continental," and in the "snow abstract annihilation" at the end of his life in "Snowed Up." Individual actions ripple out to symbolically imply things about ever greater structures that encompass (like an outer ring in a concentric pattern) the structure within it. This signature structure may owe something to Edward Carpenter's comment that around humanity, "all creation … ranges itself, as it were, like some vast flower in concentric circles" (qtd in Delavenay, *Edwardian Transition* 76).

26 As numerous commentators have observed, the "rubbing" scene in *Aaron's Rod* is comparable in a number of ways to the scene in *The White Peacock* where George rubs Cyril dry after they have bathed in the mill-pond. The element of illness is lacking in *The White Peacock* scene, however. And while George (and Annable, the gamekeeper) are father substitutes for Cyril, among other things (Cyril is kept ignorant of his father's identity until his father dies), Lilly in *Aaron's Rod* consciously takes on a maternal role here: "'I'm going to rub you as mothers do their babies whose bowels don't work'" (*AR* 96). Cowan argues

optimistically that "[t]he mother and baby analogy has far less to do with authoritarian power … than with nurturance" ("Touch" 127). Lilly's action would, then, be ironic in the Lawrentian sense, since he is acting from his sympathetic centre to awaken Aaron's volitional zone – for the lower back side of the body is, in Lawrence's scheme, the centre of volition (*FU* 243–4). Lilly is trying to activate Aaron's centre of differentiation because Aaron's illness is due to his having "given himself away." The constipation is a symptom of his failure to expel or oppose the not-self. What is most interesting is that the misogyny in *Aaron's Rod* is of a very peculiar kind. Lilly is housewifely and maternal, at least when he chooses to be, and on several occasions he accuses Aaron of acting or speaking like a woman (*AR* 105, 290). It almost seems as if feminine traits are considered too good for women. In fact, Hilary Simpson has argued that *Aaron's Rod* "is a kind of mirror-image of the traditional feminist novel" (109).

27 Mark Kinkead-Weekes speculates about a much earlier origin to the horror of confinement evident in Lawrence's work. Discussing "Education of the People," Kinkead-Weekes suggests that Lawrence had a sense by this time that his childhood had gone seriously wrong in that his mother made him conscious far too early, thereby creating "a young man whose self- and mental-consciousness became a prison that only the 'death' of sexual fulfilment could crack open" (591). It is interesting to see how in *Aaron's Rod* heterosexual relations have now become part of the prison enclosure rather than the key to unlock the prison door. Katherine Hayles suggests an urge to break through confinement in Lawrence's very ideal of balancing opposites in a tense instability that could "break through to an unbounded space that encompasses all opposites," a space that Hayles identifies as the unconscious (89). That breakthrough once achieved, however, the individual is faced with an identification of that state with "the child's fusion with the mother" (90). Hayles, then, sees paradox as the essence, even the engine, of Lawrence's art. Conceivably, one characteristic that might help to explain the problems so many have had with Lawrence's "leadership" fiction is a degeneration of pure paradox into mere contradiction.

28 It may well be that spatialization in modernist fiction bespeaks a "flattening out" of literature analogous to developments in the plastic arts toward abstraction and figure-ground conflation, but what twentieth-century fiction loses in historical "depth" it clearly gains in the exploration of a psychological dimension. The dimension of depth, in effect, becomes the psyche portrayed through time. A work like Woolf's *Mrs Dalloway* (1925) shows that consciousness is as much a time as a space; memory, as in Proust, is a virtual dimension of lived experience.

29 Balbert suggests in connection with *The Lost Girl* that Alvina searches constantly "for a man who will not wilt" (386). That search is also evident in *Aaron's Rod*, but the searcher is now male.

30 The internal quotations in Gilman are from Anatole Leroy-Beaulieu, *Israel Among the Nations: A Study of the Jews and Antisemitism*, tr. Frances Hellman (New York, 1985), a work first published in Paris in 1893. In *The Captain's Doll*, the narrator's description of the Jews encountered by Hepburn and Hannele in the Alpine hotel where they stay on their excursion to the glacier is very much in keeping with such a sense of ego vulnerability. These Jews are people who, on one hand, are condemned for pretending to be something they are not – "so that you might think they *were* Austrian aristocrats, if you weren't properly listening, or if you didn't look twice" – but, on the other hand, are appreciated somehow, for "they imparted a wholesome breath of sanity, disillusion, unsentimentality to the excited 'Bergheil' atmosphere" (*Fox* 140). The most crucial context for the condemnation of the Austrian Jews' efforts at assimilation in *The Captain's Doll* is the importance of maintaining rigidly demarcated boundaries of all sorts. In this sense, then, Lawrence's anti-Semitic attitude may be seen as a displacement of a fear of amorphous boundaries, ultimately those protecting his vulnerable sense of self. For a fuller discussion of this issue, see my "'Jews of the Wrong Sort': D.H. Lawrence and Race."

31 Kessler maintains that Lawrence "is stonily serious about race; for without differentiation according to race, there is no differentiation of blood, and without blood differences no distinctions can be possible between the 'gods' of the blood, hence no innate superiority exists, no absolutes by which to gauge inferiority, no means for establishing servile races"(484).

32 In Conrad's novel, the explosion is political but not in the way it is made to appear. Mr Verloc is a secret agent working for the Russian embassy in Tsarist days (and doubling as an informer for the British police). He is instructed by the Russians to plant a bomb at the Greenwich Observatory in order to foment a public outcry against Tsarist enemies in London. Verloc uses his mentally challenged brother-in-law, Stevie, for the job, but Stevie succeeds in destroying only himself.

33 I am thinking here of works such as Angela Carter's surrealistic *The Infernal Desire Machines of Dr Hoffman*, Russell Hoban's futurist *Riddley Walker*, and Susan Musgrave's chilling *The Charcoal Burners*.

34 Katherine Mansfield, in a May 1916 letter to S.S. Koteliansky, reported that when Lawrence "is in a rage with Frieda he says it is she who has done this to him and that she is 'a bug who has fed on my life'" (263).

35 Virginia Hyde sees the cannibalism motif in a cultural context: "The death-committed society eats the effigy of its own lack of essence"

("Revisionist Typology" 478). Paul Gleason suggests the intriguing idea that in *Aaron's Rod* Lawrence is critiquing Platonic philosophy rather directly and that Lilly and Aaron can be associated with Socrates and Alcibiades respectively (321–2). There may be something to this idea, but Gleason's further argument that "[w]ith Platonic love as the sickness, Lilly attempts to cure Aaron, not through Socratic dialectic, but through homoerotic touch" (322) surely overstates the case. While love in all of its manifestations, sexual as well as cultural, is denounced in the transitional and leadership works, it seems to me clear that it is Aaron's physical love with Josephine first and then with the Marchesa that leads directly to his serious physical malaise.

36 Howe finds it ironic "that a novel so marred by disease should take for its center, both literal and figurative, the physically flawless young male body, Michelangelo's David" (89).

37 Lawrence told his American publisher, Thomas Seltzer, that *Aaron's Rod* "is the last of my serious English novels – the end of *The Rainbow, Women in Love* line" (*L* iv 92).

38 In this letter from February 1916 Lawrence, astonishingly, asks Russell to leave him some money in his will. Given Lawrence's use of the infant/child metaphor in *Aaron's Rod,* it is also interesting to note in this letter Lawrence urging Russell to "start at the very beginning and be a perfect baby" (*L* ii 547).

39 John Turner writes that "[t]he real *spectator ab extra* in the novel … is also its creator: Lawrence himself in his role as novelist. It is he who inhabits the ultimate distances of his texts and occupies its final perspectives" ("Comedy" 79).

40 As Baker suggests, *Aaron's Rod* "represents a conscious attempt on the novelist's part to turn away from the aesthetic form of the novel which he had so successfully achieved in *The Rainbow* and *Women in Love* towards a newer kind of fiction" (vii).

CHAPTER FOUR

1 There also seems to be a question about John Thomas's gender identification, despite his Lothario reputation. When entering the women conductors' waiting room, he is greeted with "Ladies only" and responds "That's me!" In addition, of the popular songs of the era referred to in the tale, one was popularized by a male impersonator of the time (*EmyE* 40, 238n). John Thomas's sexual predatoriness may, indeed, be related as much to his own doubts about his sexuality as to the ready availability of single women in close proximity. Lawrence was no sexual predator, but the doubts about gender identification that are just barely hinted at in the story may be part of the biographical material that

seeps into the narrative, as the attack on John Thomas by the women certainly is.

2 Harris suggests a connection between Lawrence's "insistance [*sic*] on choice through the war tales" of *England, My England* and his opinion "that the English people have chosen this war, chosen to die" (116).

3 Kinkead-Weekes argues that, when the story was revised, it became "more complex, with a new sense of hurt and damage in its apparently virago heroine" (686).

4 This essay appeared originally in *Literature and History* 7 (1981): 210–22.

5 Paul Delany suggests, à propos of "Tickets Please," that Lawrence "found something particularly sinister in women controlling machinery and taking over traditionally masculine jobs" (*Nightmare* 282).

6 Cushman points out, for example, that the blindness in "The Blind Man" finds an echo in "The Horse-Dealer's Daughter" when the doctor answers Mabel's hope for love "blindly," and that Pervin is the surname of the principal characters in both of these stories. Duane Smith goes even further, and suggests that the volume might be seen as a "fragmentary novel" like *Winesburg, Ohio* or *In Our Time* (247).

7 Cushman similarly remarks that in "the most sardonic" tales in the volume (he names "The Blind Man," "Tickets Please," "Monkey Nuts," "You Touched Me," and "Wintry Peacock"), Lawrence "seems almost to be playing games with his themes" (499).

8 At the very end of the story, there seems to be another projection of the poison motif, when Berry says to himself of his uncle and his uncle's latest, much younger paramour: "'She'll hate him like poison. And serve him right'" (*EmyE* 136).

9 In *Aaron's Rod*, the first lightning description is clearly associated with reclaimed masculinity: "[Aaron] moved about in the splendour of his own male lightning, invested in the thunder of the male passion-power. He had got it back, the male godliness, the male godhead" (*AR* 258). Shortly afterwards, however, as if to make it clear that *any* heterosexual encounter at this point is debilitating to the male, Lawrence's narrator suggests that the lightning Aaron feels surging through his body in his encounter with the Marchesa is damaging in the long run: "He felt curiously blazed, as if some flame or electric power had gone through him and withered his vital tissue" (*AR* 263). In a recent article, Andrew Harrison has demonstrated a link between Lawrence's vocabulary of electricity (especially in *Women in Love*) and his reading of the Italian Futurists.

10 Ruderman likens Egbert to Maurice in "The Blind Man" in terms of dependency on a woman: "the main character is a lost boy dependent on his mother-wife, and his feeling of isolation is exacerbated by the fact that there are other, 'real' children demanding her attention" (87).

11 We can compare Widmer on this point, who writes that "Lawrence must be understood as at least partly his own Georgian hero of ineffectual sensitivity demanding universal destruction" (21).

12 The image of Egbert the gardener precedes the war in the chronology of the story but not, of course, Lawrence's own experience of the war.

13 The event was based on what happened to the granddaughter of the Meynells, acquaintances of the Lawrences.

14 The 1915 version of the story makes use of a leg motif as well, but not as effectively as the revised story. In the *English Review* variant, Joyce is injured in the leg, but there is, significantly, no snake-eating-frog episode. Instead there is an emphasis, at the end of the tale, on a serious leg wound sustained by Evelyn (as Egbert is called there) in addition to a head wound. He decides in his soul that he will die, but it is not clear if the injuries on their own would have been fatal. In a rather mawkish episode that Lawrence wisely cut from the revised story, Evelyn shoots and kills numerous German soldiers who have come to investigate the damage their barrage has created before he himself is killed and mutilated by one of their comrades-in-arms. There is no leg wound to Egbert in the 1922 version.

15 In *The Crown*, Lawrence recounts the real-life episode that inspired the snake-frog scene in "England, My England." The fictional description is faithful to what appears to have occurred in life. Lawrence's considered attitude in the essay to the episode reveals both an ambivalence regarding the natural struggle for survival as he witnessed it and his belief that the dislocated state of the vital hierarchy reflects the state of the world: "We were all white with fear. But why? In the world of twilight as in the world of light, one beast shall devour another. The world of corruption has its stages, where the lower shall devour the higher, *ad infinitum*" (*PII* 408). Snakes have always eaten frogs, of course, but it is the symbolic significance of the incident that troubled Lawrence and resonates in the fiction.

16 As indicated above in chapter 2, there are a good many soldiers in Lawrence's writing, and some have speculated about a peculiar attraction Lawrence might have had for men in khaki. There is the officer in "The Mortal Coil" and the orderly in "The Prussian Officer," to name two from his early writing. In *The Fox*, Lawrence makes it clear that Henry is not really a soldier but a hunter. Hepburn gives up his commission in *The Captain's Doll* before he successfully woos Hannele. Count Dionys in *The Ladybird* is more a prisoner than a soldier.

17 The original manuscript came to light after the Cambridge University Press edition of *England, My England* was published based on the revised version (Kinkead-Weekes 841n14).

18 Mark Spilka's view that Maurice, by the end, "moves toward greater fullness of being" and transcends his blindness ("Ritual Form" 115) is too optimistic a reading. Cushman goes so far as to say that Spilka, "impressed by the presentation of this world," is fooled into thinking that Maurice benefits from the encounter (497).

19 Tate suggests that the story is typical of its period in that it "issues a number of uncompromising but contradictory statements which cannot be resolved" (105).

20 Compare Delany: "this realm of light is Bertie's territory and Maurice chooses to withdraw from it into the darkness of the stables" ("Message and Code" 29). Delany also points out that when Bertie finds Maurice in the stable, the latter is busy pulping turnips, in Delany's words, "crushing … head-shaped objects" ("Message and Code" 29), an action that symbolizes Maurice's general hostility to an intellectual orientation and foreshadows the outcome of the scene of touching between the two men.

21 Cushman discusses John Vickery's observation that "England, My England" is patterned after the function of the scapegoat (495). Vickery's comment can be found on page 306 of his *The Literary Impact of "The Golden Bough"* (Princeton: Princeton University Press, 1973). Wheeler also suggests a concept similar to scapegoating through a laying on of hands in "The Blind Man": "The earlier cancellation of Maurice's manhood because of Bertie's presence is here transferred to the other man" ("Intimacy" 246).

22 The best and fullest argument for Bertrand Russell is Ross's "The Mythology of Friendship: D.H. Lawrence, Bertrand Russell, and 'The Blind Man.'" Ross, intriguingly, hints at the metafictional possibilities of the story: "It also represents an attempt by Lawrence to arrive at a separate armistice for a battle that, since early in the war, had been raging within his own mind" (285). Bertie, argues Ross, is not necessarily like the real Bertrand Russell but is rather "a Russell transformed by Lawrence's resentment and frustration" because of the unfortunate turn in their friendship and planned anti-war collaboration (293). Maurice is the transformed Lawrence figure, and Ross carefully traces a number of key metaphors in Lawrence's letters that find their way into the story. Paul Delany argues forcefully for J.M. Barrie as the model for Bertie (the Scottish ancestry, the fear of women) in his article "Who Was 'The Blind Man'?". John Bayley states flatly that "[t]he character of Bertie is based on that of J.M. Barrie, the playwright, whom Lawrence met through Lady Cynthia Asquith, and who combined great kindness and sympathy with a horror of any sort of physical intimacy. He was also a very short man" (125–6). W.S. Marks III, among others, has nominated Murry as Bertie's original.

23 The letter can be found in *L* ii 415. Barbara Schapiro relates Bertie to Lawrence himself and to Dollie in "The Princess" in terms of hollowness at the core of the self (65). Given the centrality of eyes in the story, we might also associate Bertie with Gerald Crich in *Women in Love*, a man who, once he has subjugated matter to his will in the coal industry, imagines that his own face is unreal and that his eyes are blue bubbles that may burst at any time "and leave clear annihilation" (*WL* 232).

24 Tate's cultural studies approach to the story suggests a textual rather than a biographical identification with both Maurice and Bertie within the context of a powerful homoerotic drive that "at the same time … tries to suppress its own queer interests" (108).

25 Schapiro writes: "The scene depicts … a collapsed intersubjectivity and the resulting dynamic of domination and submission … [T]he empty character desires in the other precisely what he himself lacks. His own deficiency and shame, however, pollute that desire and provoke the fantasy of annihilation" (65).

26 Regina Fadiman, whose reading of the story compares it to a classical ballet in structure, places Isabel "in the center of the stage" (61).

27 Judith Puchner Breen argues that Isabel needs a surrogate for herself as maternal mate to Maurice because of her desire to devote herself to the child about to be born. She chooses Bertie for the role and sends him out to Maurice. Although things do not turn out as planned and Maurice "is left with a second sham marriage," Isabel "views with complacent indifference the consequences of the trap she set for Bertie, gladly sacrificing both husband and friend to the new life within her womb" (68).

28 In Breen's words, "The impending arrival of Bertie alerts the reader to the mixed reception awaiting the other expected newcomer, the new baby" (65). Marks writes that "Bertie's rather foetal physique instantly betrays him as a kind of infantile regressive" (182).

29 In Kingsley Widmer's words, "Lawrence's erethisia, which provides the main trope in this story, appears in much of his writing" (232n19).

30 See, for example, Ruderman 81–2 and Harris 151.

31 In addition to "Hadrian," Cowan's article deals with other works by Lawrence, essays, poems, fiction, and the play *The Widowing of Mrs Holroyd*. The article became the basis of one chapter in Cowan's subsequent book, *D.H. Lawrence and the Trembling Balance* (1990), where the analysis of "Hadrian" is basically unchanged from the article. We can compare Harris's judgment: "The touch, so unlooked for, so new to this loner's experience, goes deep … [H]ere again Lawrence focuses on a literal laying on of hands, a touch, to dramatize a contract or connection between people that extends fathoms below their daytime, conscious selves" (151). Kathryn A. Walterscheid has published *The*

Resurrection of the Body: Touch in D.H. Lawrence (1993), but, oddly, the study does not deal with "Hadrian."

32 In a more recent essay, Kenneth Mackenzie largely agrees with Leavis: "Ted Rockley's apparent 'malevolence,' below the level of such 'human' feelings, is creative in effect ... there is such a thing as being 'cruel in order to be kind'" (135). He also avers that the final scene of the story "conveys a sense of reconciliation, as of a regal handing on or over" (136).

33 The letter can be found in *L* iii 475. Ruderman's view of Lawrence's identification with the two male characters in the story is that "Lawrence splits himself between the small but vigorous charity boy, desirous of taking revenge on those women who have nurtured and 'civilized' him, and the sickly father, cared for by his daughters" (82).

34 The editors of the Cambridge University Press edition of *England, My England* inform us that Lawrence's preferred new title for the story, "Hadrian," "was ignored by [his publisher] Seltzer, who retained the old title" (xlii).

35 Such a significance is not precluded by the biographical accuracy of Lawrence's writing here, since he always transformed the raw material of life according to his own purposes. In this case, it appears that Lawrence based the Pervins on an Eastwood family known to him, as the notes in the Cambridge University Press edition of the volume tell us (*EmyE* 247n).

36 Jack Stewart emphasizes the erotic theme in the story and makes use of Freudian concepts and the "monomythic" structure of separation, initiation, and return as set out by Joseph Campbell in his *The Hero with a Thousand Faces*, but his analysis is largely congruent with Harris's basic assumption that this is a story of resurrection. Indeed, the monomythic quest itself is consistent with the structure of the central Christian narrative of death and resurrection.

37 In her interesting analysis of "The Horse-Dealer's Daughter" as a struggle "to love in the face of shame," Barbara Ann Schapiro relates the description of the pond as a foul-smelling, offensive place to "an unrestrained anality and profound self-contempt" that is at work in the story. The confrontation with shame means exposing abhorrent "bodily products" (76). Schapiro agrees that the rescue at the heart of the story is a mutual one. The story "concerns two characters who surrender their mental defenses and encounter one another on a primal level beneath the rational surface of their lives" (73). Stewart also speaks of "mutual regeneration" (521).

38 As Harris notes, the persuasive idea of Fergusson's rescue of Mabel as a symbolic obstetric procedure comes from Clyde de L. Ryals, who calls it "a parody, probably conscious on Lawrence's part, of the birth of a child. It is indicative of this that Fergusson is made a doctor" (156).

Ryals also mentions the pertinence to a childbirth of Fergusson's working to remove fluid from Mabel's lungs and his wrapping her up in his overcoat (156). Lawrence's narrative strategy of making Fergusson a doctor may well stem from the idea of making him a midwife of sorts to Mabel's (re)birth. However, his function as a curer of illness is, I think, also relevant. In fact, Joe Pervin, to Jack's annoyance, remarks upon the irony of a doctor ill with a cold: "'It's a knock-out, isn't it … if a doctor goes round with a cold'" (*EmyE* 140). Ryals' essay was originally published in *Literature and Psychology* in 1962.

39 Mackenzie speaks of a pervasive "sense of failure, of foiled and coarsened living" (133).

40 The Jungian reading Cushman mentions is Ryals' analysis.

41 Stewart recognizes the connection between the doctor's illness and "Lawrence with his 'bronchials'" (518), an allusion to Lawrence's habitual euphemistic reference to his tuberculosis.

42 Cushman suggests "Cinderella" and "Sleeping Beauty" as sources (495, 496) while Mackenzie compares the tale to Eliot's *Silas Marner* (133), often viewed as folkloric in form.

43 In arguing that *England, My England* may be viewed as a "fragmentary novel," Smith posits the Pluto-Persephone myth and variations of it as a structural feature common throughout the volume of stories (251–2).

CONCLUSION

1 Murry wrote: "No two imaginary characters in the whole of Lawrence's fiction are more clearly related than these. Essentially, they are the same figure" (40).

2 Ellis is referring to *The First Lady Chatterley* here, but the statement is equally true of the final version.

Works Cited

Arndal, Lisbeth. "Lawrence, Darwin and Hardy: The Influence of Evolutionary Thought on D.H. Lawrence's Fiction with Special Reference to *The Rainbow* and *Women in Love*." Diss. University of Copenhagen, 1991.

Baker, Paul G. *A Reassessment of D.H. Lawrence's "Aaron's Rod."* Ann Arbor: UMI Research Press, 1983.

Bakhtin, Mikhail. *Rabelais and His World*. Tr. Helene Iswolsky. Cambridge, MA: MIT Press, 1968.

Balbert, Peter. "Ten Men and a Sacred Prostitute: The Psychology of Sex in the Cambridge Edition of *The Lost Girl*." *Twentieth Century Literature* 36 (1990): 381–402.

Barr, William R. "*Aaron's Rod* as D.H. Lawrence's Picaresque Novel." *D.H. Lawrence Review* 9 (1976): 213–25.

Bayley, John. *The Short Story: Henry James to Elizabeth Bowen*. Brighton, Sussex: Harvester Press, 1988.

Beer, Gillian. *Darwin's Plots: Evolutionary Narrative in Darwin, George Eliot and Nineteenth-Century Fiction*. London: Routledge and Kegan Paul, 1983.

Bettelheim, Bruno. *The Uses of Enchantment: The Meaning and Importance of Fairy Tales*. New York: Random House-Vintage, 1977.

Blavatsky, H.P. *Anthropogenesis*. Vol. 2 of *The Secret Doctrine: The Synthesis of Science, Religion, and Philosophy*. 2nd ed. 3 vols. London: Theosophical Publishing, 1888.

Bonds, Diane S. *Language and the Self in D.H. Lawrence*. Ann Arbor: UMI Research Press, 1987.

Booth, Howard J. "'Give me *differences*': Lawrence, Psychoanalysis, and Race." *D.H. Lawrence Review* 27 (1997–98): 171–96.

Breen, Judith Puchner. "D.H. Lawrence, World War I and the battle between the sexes: A reading of "The Blind Man" and "Tickets Please." *Women's Studies* 13 (1986): 63–74.
Campbell, Joseph. *The Masks of God*. Vol. 1. New York: Viking, 1959.
Cavitch, David. "Merging – with Fish and Others." *D.H. Lawrence Review* 7 (1974): 172–8.
Clark, L.D. "Making the Classic Contemporary: Lawrence's Pilgrimage Novels and American Romance." In *D.H. Lawrence in the Modern World*. Ed. Peter Preston and Peter Hoare. 193–216. Cambridge: Cambridge University Press, 1989.
Conrad, Joseph. *The Secret Agent: A Simple Tale*. 1907. Ed. Martin Seymour-Smith. Harmondsworth: Penguin, 1984.
Cowan, James C. *D.H. Lawrence and the Trembling Balance*. University Park, PA: Pennsylvania State University Press, 1990.
– "Lawrence and Touch." *D.H. Lawrence Review* 18 (1985–86): 121–37.
Cushman, Keith. "The Achievement of *'England My England' and Other Stories*. In *D.H. Lawrence: Critical Assessments*. Ed. David Ellis and Ornella De Zordo. 4 vols. 3:490–502. Mountfield: Helm Information, 1992.
Daleski, H.M. "Aphrodite of the Foam and *The Ladybird* Tales." In *D.H. Lawrence: A Critical Study of the Major Novels and Other Writings*. Ed. A.H. Gomme. 142–58. Hassocks, UK: Harvester Press; New York: Barnes and Noble, 1978.
– *The Forked Flame: A Study of D.H. Lawrence*. London: Faber, 1965.
Darroch, Robert. *D.H. Lawrence in Australia*. Melbourne: Macmillan, 1981.
Darwin, Charles. *The Origin of Species by Means of Natural Selection or the Preservation of Favoured Races in the Struggle for Life*. 1859. New York: Mentor-New American Library, 1958.
Davis, Patricia C. "Chicken Queen's Delight: D.H. Lawrence's 'The Fox.'" *Modern Fiction Studies* 19 (1973–74): 565–71.
Dawson, Eugene W. "Love Among the Mannikins: *The Captain's Doll*." *D.H. Lawrence Review* 1 (1968): 137–48.
Delany, Paul. *D.H. Lawrence's Nightmare: The Writer and His Circle in the Years of the Great War*. New York: Basic Books, 1978.
– "'We shall know each other now': Message and Code in D.H. Lawrence's 'The Blind Man.'" *Contemporary Literature* 26 (1985): 26–39.
– "Who Was 'The Blind Man'?" In *D.H. Lawrence: Critical Assessments*. Ed. David Ellis and Ornella De Zordo. 4 vols. 3:526–32. Mountfield: Helm Information, 1992.
Delavenay, Emile. *D.H. Lawrence and Edward Carpenter: A Study in Edwardian Transition*. London: Heinemann, 1971.
– *D.H. Lawrence: The Man and his Work. The Formative Years: 1885–1919*. Tr. Katherine Delavenay. London: Heinemann, 1972.
– "Further Thoughts on Lawrence and Chamberlain." *D.H. Lawrence Review* 19 (1987): 173–80.

Devlin, Albert J. "The 'strange and fiery' course of *The Fox*: D.H. Lawrence's Aesthetic of Composition and Revision." In *The Spirit of D.H. Lawrence: Centenary Studies*. Ed. Gamini Salgado and G.K. Das. 75–91. Totowa, NJ: Barnes and Noble, 1988.

Doherty, Gerald. *Theorizing Lawrence: Nine Meditations on Tropological Themes*. New York: Peter Lang, 1999.

– "A 'very funny' story: Figural Play in D.H. Lawrence's *The Captain's Doll*." *D.H. Lawrence Review* 18 (1985–86): 5–17.

Ebbatson, Roger. *The Evolutionary Self: Hardy, Forster, Lawrence*. Brighton, Sussex: Harvester Press, 1982.

– *Lawrence and the Nature Tradition: A Theme in English Fiction 1859–1914*. Brighton, Sussex: Harvester Press, 1980.

– "A Spark Beneath the Wheel: Lawrence and Evolutionary Thought." In *D.H. Lawrence: New Studies*. Ed. Christopher Heywood. 90–103. New York: St Martin's Press, 1987.

Edge, John. "D.H. Lawrence and the Theme of Comradeship." *Southern Review* (Adelaide) 9.1 (1976): 34–49.

Eliot, T.S. *After Strange Gods: A Primer of Modern Heresy*. London: Faber, 1934.

Ellis, David. *D.H. Lawrence: Dying Game, 1922–1930*. Cambridge: Cambridge University Press, 1998.

Engel, Monroe. "The Continuity of Lawrence's Short Novels." In *D.H. Lawrence: A Collection of Critical Essays*. Ed. Mark Spilka. 93–100. Englewood Cliffs, NJ: Prentice-Hall, 1963.

Fadiman, Regina. "The Poet as Choreographer: Lawrence's 'The Blind Man.'" *Journal of Narrative Technique* 2 (1972): 60–7.

Fernihough, Anne. *D.H. Lawrence: Aesthetics and Ideology*. Oxford: Clarendon Press, 1993.

Ford, George H. *Double Measure: A Study of the Novels and Stories of D.H. Lawrence*. New York: Holt, Rinehart and Winston, 1965.

Forster, E.M. *Howards End*. 1910. Ed. Oliver Stallybrass. London: Penguin, 1989.

Foster, John Burt, Jr. *Heirs to Dionysus: A Nietzschean Current in Literary Modernism*. Princeton: Princeton University Press, 1981.

Fowler, Roger. "*The Lost Girl*: discourse and focalization." In *Rethinking Lawrence*. Ed. Keith Brown. 53–66. Milton Keynes: Open University Press, 1990.

Gilbert, Sandra M. "Potent Griselda: 'The Ladybird' and the Great Mother." In *D.H. Lawrence: A Centenary Consideration*. Ed. Peter Balbert and Phillip L. Marcus. 130–61. Ithaca: Cornell University Press, 1985.

Gilman, Sander L. *Picturing Health and Illness: Images of Identity and Difference*. Baltimore: Johns Hopkins University Press, 1995.

Gleason, Paul. "A Note on Plato and *Aaron's Rod*." *D.H. Lawrence Review* 27 (1997–98): 321–4.

Good, Jan. "Toward a Resolution of Gender Identity Confusion: The Relationship of Henry and March in *The Fox*." *D.H. Lawrence Review* 18 (1985–86): 217–27.

Granofsky, Ronald. "'Jews of the Wrong Sort': D.H. Lawrence and Race." *Journal of Modern Literature* 23 (1999–2000): 209–23.

– "A Second Caveat: D.H. Lawrence's *The Fox*." *English Studies in Canada* 14 (1988): 49–63.

Grimm, the Brothers. *The Complete Fairy Tales of the Brothers Grimm*. Tr. Jack Zipes. Toronto: Bantam, 1987.

Gurko, Leo. "D.H. Lawrence's Greatest Collection of Short Stories: What Holds It Together." *Modern Fiction Studies* 18 (1972–73): 173–82.

Harris, Janice Hubbard. *The Short Fiction of D.H. Lawrence*. New Brunswick, NJ: Rutgers University Press, 1984.

Harrison, Andrew. "Electricity and the Place of Futurism in *Women in Love*." *D.H. Lawrence Review* 29 (2000): 7–23.

Hayles, N. Katherine. *The Cosmic Web: Scientific Field Models and Literary Strategies in the Twentieth Century*. Ithaca: Cornell University Press, 1984.

Herring, Phillip. "Caliban in Nottingham: D.H. Lawrence's *The Lost Girl*." *Mosaic* 12.4 (1979): 9–19.

Herzinger, Kim A. *D.H. Lawrence in His Time: 1908–1915*. Lewisburg: Bucknell University Press; London: Associated University Presses, 1982.

Heywood, Christopher. "*Birds, Beasts and Flowers*: The Evolutionary Context and Lawrence's African Literary Source." *D.H. Lawrence Review* 15 (1982): 87–105.

Holbrook, David. *Where D.H. Lawrence Was Wrong about Woman*. Lewisburg: Bucknell University Press; London: Associated University Presses, 1992.

Holderness, Graham. *D.H. Lawrence: History, Ideology and Fiction*. Dublin: Gill and Macmillan, 1982.

Hough, Graham. *The Dark Sun: A Study of D.H. Lawrence*. Duckworth: London, 1956.

Howarth, Herbert. "D.H. Lawrence from Island to Glacier." *University of Toronto Quarterly* 37 (1967–68): 215–29.

Howe, Marguerite Beede. *The Art of the Self in D.H. Lawrence*. Athens, OH: Ohio University Press, 1977.

Huxley, Aldous. "Introduction." *The Letters of D.H. Lawrence*. ix-xxxiv. London: Heinemann, 1932.

Hyde, G.M. *D.H. Lawrence*. New York: St Martin's Press, 1990.

Hyde, Virginia. "*Aaron's Rod*: D.H. Lawrence's Revisionist Typology." In *D.H. Lawrence: Critical Assessments*. Ed. David Ellis and Ornella De Zordo. 4 vols. 2:468–82. Mountfield: Helm Information, 1992.

– "'Lost' Girls: D.H. Lawrence's Versions of Persephone." In *Images of Persephone: Feminist Readings in Western Literature*. Ed. Elizabeth T. Hayes. 99–120. Gainesville, FL: University Press of Florida, 1994.

Jones, Lawrence. "Physiognomy and the Sensual Will in *The Ladybird* and *The Fox*." *D.H. Lawrence Review* 13 (1980): 1–29.

Kern, Stephen. *The Culture of Time and Space 1880–1918*. Cambridge, MA: Harvard University Press, 1983.

Kessler, Jascha. "D.H. Lawrence's Primitivism." *Texas Studies in Literature and Language* 5 (1964): 467–88.

Kiely, Robert. "Accident and Purpose: 'Bad Form' in Lawrence's Fiction." In *D.H. Lawrence: A Centenary Consideration*. Ed. Peter Balbert and Phillip L. Marcus. 91–107. Ithaca: Cornell University Press, 1985.

Kinkead-Weekes, Mark. *D.H. Lawrence: Triumph to Exile 1912–1922*. Cambridge: Cambridge University Press, 1996.

Krasner, James. *The Entangled Eye: Visual Perception and the Representation of Nature in Post-Darwinian Narrative*. New York: Oxford University Press, 1992.

Lawrence, D.H. *Aaron's Rod*. 1922. Ed. Mara Kalnins. Cambridge: Cambridge University Press, 1988.

– *The Complete Poems of D.H. Lawrence*. Ed. Vivian de Sola Pinto and Warren Roberts. 1964. New York: Penguin, 1977.

– *The Complete Short Stories of D.H. Lawrence*. Vol. 1. London: Heinemann, 1955.

– *D.H. Lawrence: Three Plays*. London: Penguin, 1969.

– *England, My England and Other Stories*. 1922. Ed. Bruce Steele. Cambridge: Cambridge University Press, 1990.

– *Fantasia of the Unconscious and Psychoanalysis and the Unconscious*. Harmondsworth: Penguin, 1971.

– "The Fox." In *A D.H. Lawrence Miscellany*. Ed. Harry T. Moore. 28–48. Carbondale: Southern Illinois University Press, 1959.

– *The Fox, The Captain's Doll, The Ladybird*. Ed. Dieter Mehl. Cambridge: Cambridge University Press, 1992.

– *The Letters of D.H. Lawrence: Volume I: September 1901-May 1913*. Ed. James T. Boulton. Cambridge: Cambridge University Press, 1979.

– *The Letters of D.H. Lawrence: Volume II: June 1913-October 1916*. Ed. George J. Zytaruk and James T. Boulton. Cambridge: Cambridge University Press, 1981.

– *The Letters of D.H. Lawrence: Volume III: October 1916-June 1921*. Ed. James T. Boulton and Andrew Robertson. Cambridge: Cambridge University Press, 1984.

– *The Letters of D.H. Lawrence: Volume IV: June 1921-March 1924*. Ed. Warren Roberts, James T. Boulton, and Elizabeth Mansfield. Cambridge: Cambridge University Press, 1987.

– *The Letters of D.H. Lawrence: Volume VI: March 1927-November 1928*. Ed. James T. Boulton and Margaret H. Boulton, with Gerald M. Lacy. Cambridge: Cambridge University Press, 1991.

– *The Letters of D.H. Lawrence: Volume VII: November 1928-February 1930.* Ed. Keith Sagar and James T. Boulton. Cambridge: Cambridge University Press, 1993.
– *The Lost Girl.* 1920. Ed. John Worthen. Cambridge: Cambridge University Press, 1981.
– *Mornings in Mexico and Etruscan Places.* Harmondsworth: Penguin, 1974.
– *Movements in European History.* 1921. Ed. Philip Crumpton. Cambridge: Cambridge University Press, 1989.
– *Mr Noon.* 1984. Ed. Lindeth Vasey. London: Grafton-Collins, 1986.
– *Phoenix: The Posthumous Papers of D.H. Lawrence.* Ed. Edward D. McDonnald. 1936. Harmondsworth: Penguin, 1978.
– *Phoenix II: Uncollected, Unpublished and Other Prose Works by D.H. Lawrence.* Ed. Warren Roberts and Harry T. Moore. 1968. New York: Viking, 1968.
– *The Rainbow.* 1915. Ed. Mark Kinkead-Weekes. Cambridge: Cambridge University Press, 1989.
– *Studies in Classic American Literature.* 1924. Harmondsworth: Penguin, 1971.
– *The Symbolic Meaning: The Uncollected Versions of "Studies in Classic American Literature."* Ed. Armin Arnold. London: Centaur Press, 1962.
– *The White Peacock.* 1911. Ed. Andrew Robertson. London: Penguin, 1995.
– *Women in Love.* 1920. Ed. David Farmer, Lindeth Vasey, and John Worthen. Cambridge: Cambridge University Press, 1987.
Lawrence, Frieda. *Not I, But the Wind.* Toronto: Macmillan, 1934.
Leavis, F.R. *D.H. Lawrence: Novelist.* 1955. Harmondsworth: Pelican-Penguin, 1973.
– *Thought, Words and Creativity: Art and Thought in Lawrence.* London: Chatto, 1976.
Levenson, Michael H. *A Genealogy of Modernism: A Study of English Literary Doctrine 1908–1922.* Cambridge: Cambridge University Press, 1984.
Levine, George. *Darwin and the Novelists: Patterns of Science in Victorian Fiction.* 1988. Chicago: University of Chicago Press, 1991.
Lüthi, Max. *Once Upon a Time: On the Nature of Fairy Tales.* Tr. Lee Chadeayne and Paul Gottwald with additions by the author. Bloomington: Indiana University Press, 1976.
MacDonald, Robert H. "'The Two Principles': A Theory of the Sexual and Psychological Symbolism of D.H. Lawrence's Later Fiction." *D.H. Lawrence Review* 11 (1978): 132–55.
McDowell, Frederick P.W. "'The Individual in His Pure Singleness': Theme and Symbol in *The Captain's Doll.*" In *The Challenge of D.H. Lawrence.* Ed. Michael Squires and Keith Cushman. 143–58. Madison: University of Wisconson Press, 1990.
Mackenzie, Kenneth. "Energy and Ennui in *England, My England.*" In *D.H. Lawrence: A Critical Study of the Major Novels and Other Writings.*

Ed. A.H. Gomme. 120–41. Hassocks, UK: Harvester Press; New York: Barnes and Noble, 1978.

Mansfield, Katherine. *The Collected Letters of Katherine Mansfield.* Ed. Vincent O'Sullivan and Margaret Scott. 3 vols. Vol. 1: 1903–1917. Oxford: Clarendon Press, 1984.

Marks, W.S. III. "The Psychology of Regression in D.H. Lawrence's 'The Blind Man.'" *Literature and Psychology* 17 (1967): 177–92.

Mellown, Elgin W. "*The Captain's Doll*: Its Origins and Literary Allusions." *D.H. Lawrence Review* 9 (1976): 226–35.

Mensch, Barbara. *D.H. Lawrence and the Authoritarian Personality.* New York: St Martin's Press, 1991.

Meyers, Jeffrey. *D.H. Lawrence: A Biography.* New York: Vintage-Random House, 1992.

– *D.H. Lawrence and the Experience of Italy.* Philadelphia: University of Pennsylvania Press, 1982.

Moore, Harry T. *The Intelligent Heart: The Story of D.H. Lawrence.* New York: Farrar, Strauss, 1954.

Moynahan, Julian. *The Deed of Life: The Novels and Tales of D.H. Lawrence.* Princeton: Princeton University Press, 1963.

Murry, J. Middleton. *D.H. Lawrence: Son of Woman.* 1931. London: Jonathan Cape, 1972.

Nehls, Edward, ed. *D.H. Lawrence: A Composite Biography.* 3 vols. Madison: University Wisconsin Press, 1957–59.

Nelson, Jane. "The Familial Isotopy in *The Fox*." In *The Challenge of D.H. Lawrence.* Ed. Michael Squires and Keith Cushman. 129–42. University Wisconsin Press, 1990.

Norris, Margot. *Beasts of the Modern Imagination: Darwin, Nietzsche, Kafka, Ernst, and Lawrence.* Baltimore: Johns Hopkins University Press, 1985.

Paradis, James and George C. Williams. *Evolution and Ethics: T.H. Huxley's "Evolution and Ethics" with New Essays on its Victorian and Sociobiological Context.* Princeton: Princeton University Press, 1989.

Pecora, Vincent P. "D.H. Lawrence." In *The Columbia History of the British Novel.* Ed. John Richetti. 715–39. New York: Columbia University Press, 1994.

Phillips, Steven R. "The Double Pattern of D.H. Lawrence's 'The Horse Dealer's Daughter.'" *Studies in Short Fiction* 10 (1973): 94–7.

Pinkney, Tony, *D.H. Lawrence and Modernism.* Iowa City: University of Iowa Press, 1990.

Ross, Michael L. "The Mythology of Friendship: D.H. Lawrence, Bertrand Russell, and 'The Blind Man.'" In *English Literature and British Philosophy.* Ed. S.P. Rosenbaum. 285–315. Chicago: University of Chicago Press, 1971.

– *Storied Cities: Literary Imaginings of Florence, Venice, and Rome.* Westport, CT: Greenwood Press, 1994.

Rossi, Patrizio. "Lawrence's Two *Foxes*: A Comparison of the Texts." *Essays in Criticism* 22 (1972): 265–78.

Ruderman, Judith. *D.H. Lawrence and the Devouring Mother: The Search for a Patriarchal Ideal of Leadership*. Durham, NC: Duke University Press, 1984.

Ryals, Clyde de L. "D.H. Lawrence's 'The Horse Dealer's Daughter': An Interpretation." In *Critical Essays on D.H. Lawrence*. Ed. Dennis Jackson and Fleda Brown Jackson. 153–8. Boston: G.K. Hall, 1988.

Ryan, Kiernan. "The Revenge of the Women: Lawrence's 'Tickets Please.'" In *D.H. Lawrence: Critical Assessments*. Ed. David Ellis and Ornella De Zordo. 4 vols. 3:503–14. Mountfield: Helm Information, 1992.

Sagar, Keith. *The Art of D.H. Lawrence*. Cambridge: Cambridge UP, 1966.

Sargent, M. Elizabeth. "*The Lost Girl*: Re-appraising the Post-War Lawrence on Women's Will and Ways of Knowing." In *D.H. Lawrence in Italy and England*. Ed. George Donaldson and Mara Kalnins. 176–92. Houndmills, Basingstoke: Macmillan; New York: St Martin's Press, 1999.

Schapiro, Barbara Ann. *D.H. Lawrence and the Paradoxes of Psychic Life*. Albany: State University of New York Press, 1999.

Scheckner, Peter. *Class, Politics, and the Individual: A Study of the Major Works of D.H. Lawrence*. Rutherford: Fairleigh Dickinson University Press, 1985.

Schneider, Daniel J. *D.H. Lawrence: The Artist as Psychologist*. Lawrence, KS: University Press of Kansas, 1984.

– "D.H. Lawrence and Houston Chamberlain: Once Again." *D.H. Lawrence Review* 19 (1987): 157–71.

Scott, James F. "Thimble into Ladybird: Nietzsche, Frobenius, and Bachofen in the Later Work of D.H. Lawrence." *Arcadia* 13 (1978): 161–76.

Simpson, Hilary. *D.H. Lawrence and Feminism*. London: Croom Helm, 1982.

Smith, Duane. "*England, My England* as Fragmentary Novel." *D.H. Lawrence Review* 24 (1992): 247–55.

Spilka, Mark. *The Love Ethic of D.H. Lawrence*. Bloomington: Indiana University Press, 1955.

– "Ritual Form in 'The Blind Man.'" In *D.H. Lawrence: A Collection of Critical Essays*. Ed. Mark Spilka. 112–16. Englewood Cliffs, NJ: Prentice-Hall, 1963.

Steven, Laurence. "From Thimble to Ladybird: D.H. Lawrence's Widening Vision?" *D.H. Lawrence Review* 18 (1985–86): 239–53.

Stewart, Jack. "The Horse Dealer's Daughter." In *D.H. Lawrence: Critical Assessments*. Ed. David Ellis and Ornella De Zordo. 4 vols. 3:515–25. Mountfield: Helm Information, 1992.

Tate, Trudi. *Modernism, History and the First World War*. Manchester: Manchester University Press, 1998.

Templeton, Wayne. "D.H. Lawrence: Illness, Identity, and Writing." In *D.H. Lawrence: The Cosmic Adventure. Studies of his Ideas, Works, and*

Literary Relationships. Ed. Lawrence Gamache; assoc ed. Phyllis Sternberg Perrakis. 184–204. Nepean, ON: Borealis Press, 1996.

Thornton, Weldon. "A Trio from Lawrence's *England, My England and Other Stories*: Readings of 'Monkey Nuts,' 'The Primrose Path,' and 'Fanny and Annie.'" *D.H. Lawrence Review* 28 (1999): 5–29.

Trodd, Anthea. *A Reader's Guide to Edwardian Literature*. Calgary: University of Calgary Press, 1991.

Turner, John. "Comedy and Hysteria in *Aaron's Rod*." In *Lawrence and Comedy*. Ed. Paul Eggert and John Worthen. 70–88. Cambridge: Cambridge University Press, 1996.

– "A Punch in the Wind: Jim Bricknell Answers Back." *D.H. Lawrence Review* 28 (1999): 79–95.

Vickery, John B. "Myth and Ritual in the Shorter Fiction of D.H. Lawrence." *Modern Fiction Studies* 1 (1959): 65–82.

Vine, Steven. Introduction to *Aaron's Rod* by D.H. Lawrence. Ed. Mara Kalnins. xv-xxxvi. London: Penguin, 1995.

Vivas, Eliseo. *D.H. Lawrence: The Failure and the Triumph of Art*. Evanston, IL: Northwestern University Press, 1960.

Walterscheid, Kathryn A. *The Resurrection of the Body: Touch in D.H. Lawrence*. American University Studies IV: English Language and Literature 157. New York: Peter Lang, 1993.

Wells, H.G. *Tono-Bungay*. 1909. New York: Random House, 1938.

Wexler, Joyce. "D.H. Lawrence Through a Postmodernist Lens." *D.H. Lawrence Review* 27 (1997–98): 47–64.

Wheeler, Richard P. "'Cunning in his overthrow': Give and Take in 'Tickets Please.'" *D.H. Lawrence Review* 10 (1977): 242–50.

– "Intimacy and Irony in 'The Blind Man.'" *D.H. Lawrence Review* 9 (1976): 236–53.

Widdowson, Peter, ed. *D.H. Lawrence*. London: Longman, 1992.

Widmer, Kingsley. *The Art of Perversity: D.H. Lawrence's Shorter Fiction*. Seattle: University of Washington Press, 1962.

Wiener, Gary A. "Lawrence's 'Little Girl Lost.'" *D.H. Lawrence Review* 19 (1987): 243–53.

Williams, Raymond. "The Social Thinking of D.H. Lawrence." In *A D.H. Lawrence Miscellany*. Ed. Harry T. Moore. 295–311. Carbondale: Southern Illinois University Press, 1959.

Wollaeger, Mark A. *Joseph Conrad and the Fictions of Skepticism*. Stanford: Stanford University Press, 1990.

Worthen, John. *D.H. Lawrence: The Early Years 1885–1912*. Cambridge: Cambridge University Press, 1992.

– *D.H. Lawrence and the Idea of the Novel*. London: Macmillan, 1979.

– "Recovering *The Lost Girl*: Lost Heroines, Irrecoverable Texts, Irretrievable Landscapes." In *D.H. Lawrence in Italy and England*. Ed. George

Donaldson and Mara Kalnins. 211–27. Houndmills, Basingstoke: Macmillan; New York: St Martin's Press, 1999.

Zipes, Jack. *The Brothers Grimm: From Enchanted Forests to the Modern World.* New York: Routledge, 1988.

Zytaruk, George J. "The Doctrine of Individuality: D.H. Lawrence's 'Metaphysic.'" In *D.H. Lawrence: A Centenary Consideration.* Ed. Peter Balbert and Phillip L. Marcus. 237–53. Ithaca: Cornell University Press, 1985.

Index